American Book Company
Meeting Standards, Exceeding Expectations

Dear Educator,

Thank you for your interest in American Book Company's state-specific test preparation resources. We commend you for your interest in pursuing your students' success. Feel free to contact us with any questions about our books, software, or the ordering process.

Our Products Feature	Your Students Will Improve
Multiple-choice and open-ended diagnostic tests	Confidence and mastery of subjects
Step-by-step instruction	Concept development
Frequent practice exercises	Critical thinking
Chapter reviews	Test-taking skills
Multiple-choice practice tests	Problem-solving skills

American Book Company's writers and curriculum specialists have over 100 years of combined teaching experience, working with students from kindergarten through middle, high school, and adult education.

Our company specializes in effective test preparation books and software for high stakes graduation and grade promotion exams across the country.

How to Use This Book

Each book:

*contains a chart of standards which correlates all test questions and chapters to the state exam's standards and benchmarks as published by the state department of education. This chart is found in the front of all preview copies and in the front of all answer keys.

*begins with a full-length pretest (diagnostic test). This test not only adheres to your specific state standards, but also mirrors your state exam in weights and measures to help you assess each individual student's strengths and weaknesses.

*offers an evaluation chart. Depending on which questions the students miss, this chart points to which chapters individual students or the entire class need to review to be prepared for the exam.

*provides comprehensive review of all tested standards within the chapters. Each chapter includes engaging instruction, practice exercises, and chapter reviews to assess students' progress.

*finishes with two full-length practice tests for students to get comfortable with the exam and to assess their progress and mastery of the tested standards and benchmarks.

While we cannot guarantee success, our products are designed to provide students with the concept and skill development they need for the graduation test or grade promotion exam in their own state. We look forward to hearing from you soon.

Sincerely,

The American Book Company Team

PO Box 2638★ Woodstock, GA 30188-1383★ Phone: 1-888-264-5877★ Fax: 1-866-827-3240

BASICS MADE EASY

MATHEMATICS REVIEW

Arithmetic
Data Interpretation
Problem Solving
Algebra
Geometry

Updated November 2007

COLLEEN PINTOZZI

AMERICAN BOOK COMPANY
P O BOX 2638
WOODSTOCK, GEORGIA 30188-1383
Toll Free: 1 (888) 264-5877   Phone: 770-928-2834   Fax: 770-928-7483
Web site: www.americanbookcompany.com

ACKNOWLEDGEMENTS

Editors: Mary Reagan
Joe Wood
Kelly Berg

Copyright ©1998, 2000, 2001, 2003, 2008
by American Book Company
PO Box 2638
Woodstock, GA 30188-1383

ALL RIGHTS RESERVED

The text of this publication, or any part thereof, may not be reproduced or transmitted in any form or by any means, electronic or mechanical, including photocopying, recording, storage in an information retrieval system, or otherwise, without the prior permission of the publisher.

Printed in the United States of America
06/03   03/04   11/04   01/05   02/05   09/05   03/07   11/07

TABLE OF CONTENTS

Basics Made Easy: Mathematics Review

Preface	vi
DIAGNOSTIC EXAM	1
CHAPTER 1	**19**
Sets	
Introduction to Sets	19
Subsets	20
Intersection of Sets	21
Union of Sets	22
Chapter Review	23
Chapter Test	24
CHAPTER 2	**25**
Whole Numbers	
Adding Whole Numbers	25
Subtracting Whole Numbers	26
Multiplying Whole Numbers	28
Dividing Whole Numbers	29
Rounding Whole Numbers	30
Estimated Solutions	31
Prime and Composite Numbers	32
Time of Travel	33
Rate	34
Distance	35
Miles per Gallon	36
Word Problems	37
Chapter Review	38
Chapter Test	39
CHAPTER 3	**41**
Fractions	
Simplifying Improper Fractions	41
Mixed Numbers to Improper Fractions	42
Greatest Common Factor	43
Reducing Proper Fractions	44
Multiplying Fractions	45
Deductions - Fractions Off	46
Dividing Fractions	47
Finding Numerators	48
Ordering Fractions	49
Least Common Multiple	50
Adding Fractions	51
Subtracting Fractions	52
Fraction Word Problems	54
Chapter Review	55
Chapter Test	57

CHAPTER 4	**59**
Decimals	
Rounding Decimals	59
Reading and Writing Decimals	60
Adding Decimals	62
Subtracting Decimals	63
Determining Change	64
Multiplication of Decimals	65
Gross Pay	67
Division of Decimals by Whole Numbers	68
Changing Fractions to Decimals	69
Changing Mixed Numbers to Decimals	70
Changing Decimals to Fractions	71
Division of Decimals by Decimals	72
Best Buy	73
Ordering Decimals	75
Decimal Word Problems	76
Chapter Review	78
Chapter Test	80
CHAPTER 5	**82**
Ratios, Probability, Proportions,	
& Scale Drawings	
Ratio Problems	82
Probability	83
Solving Proportions	85
Ratio and Proportion Word Problems	86
Map and Scale Drawings	87
Chapter Review	90
Chapter Test	91
CHAPTER 6	**93**
Percents	
Changing Between Decimals and Percents	93
Changing Between Fractions and Percents	94
Representing Rational Numbers Graphically	96
Finding the Percent of the Total	97
Commissions	98
Finding the Amount of Discount	99
Finding the Discounted Sale Price	100
Sales Tax	101
Finding the Percent	102
Understanding Simple Interest	103
Buying on Credit	104
Chapter Review	105
Chapter Test	107

CHAPTER 7	**109**
Problem-Solving & Critical Thinking	
Missing Information	109
Exact Information	111
Extra Information	113
Estimated Solutions	114
Two-Step Problems	115
Pattern Problems	116
Patterns in Exchanges	118
Using Diagrams to Solve Problems	120
Trial and Error Problems	121
Chapter Review	122
Chapter Test	124
CHAPTER 8	**127**
Using Exponents & The Metric System	
Understanding Exponents	127
Square Root	128
Multiplying and Dividing by Multiples of Ten	129
The Metric System	130
Estimating Metric Measurements	132
Scientific Notation	133
Chapter Review	135
Chapter Test	136
CHAPTER 9	**138**
Customary Measurements	
Using the Ruler	138
Converting Units of Measure	140
Adding Units of Measure	141
Subtracting Units of Measure	142
Multiplying Units of Measure	143
Chapter Review	144
Chapter Test	145
CHAPTER 10	**147**
Data Interpretation	
Reading Tables	147
Mileage Chart	149
Bar Graphs	150
Line Graphs	152
Circle Graphs	154
Catalog Ordering	155
Menu Ordering	156
Chapter Review	157
Chapter Test	160

CHAPTER 11	**163**
Statistics	
Range	163
Mean	164
Median	165
Mode	166
Tally Charts and Frequency Tables	167
Histograms	169
Chapter Review	170
Chapter Test	172
CHAPTER 12	**175**
Integers & Order of Operations	
Integers	175
Absolute Value	176
Adding Integers	176
Subtracting Integers	179
Multiplying Integers	180
Mixed Integer Practice	181
Properties of Addition and Multiplication	181
Order of Operations	182
Chapter Review	183
Chapter Test	184
CHAPTER 13	**186**
Introduction to Algebra	
Algebra Vocabulary	186
Substituting Numbers for Variables	187
Understanding Algebra Word Problems	188
Setting Up Algebra Word Problems	192
Matching Algebraic Expressions	193
Chapter Review	194
Chapter Test	195
CHAPTER 14	**197**
Algebra	
Solving One-Step Algebra Problems with Addition and Subtraction	197
Solving One-Step Algebra Problems with Multiplication and Division	199
Solving One-Step Algebra Problems by Multiplying and Dividing with Negative Numbers	201
Variables with a Coefficient of Negative One	202
Two-Step Algebra Problems	203
Algebra Word Problems	204
Graphing Inequalities	205
Solving Inequalities by Addition and Subtraction	206
Solving Inequalities by Multiplication and Division	207
Chapter Review	208
Chapter Test	209

CHAPTER 15 — 211
Functions, Number Patterns, Permutations, & Combinations

Function Machines	211
Algebraic Functions	213
Function Tables	214
Identifying the Operation	215
Number Patterns	216
Other Pattern Problems	217
Permutations	219
Combinations	222
Chapter Review	224
Chapter Test	226

CHAPTER 16 — 228
Angles

Angles	228
Types of Angles	229
Measuring Angles	230
Adjacent Angles	231
Vertical Angles	232
Complementary and Supplementary Angles	233
Chapter Review	234
Chapter Test	235

CHAPTER 17 — 237
Plane Geometry

Plane Geometry Terms	237
Perimeter	238
Area of Squares and Rectangles	239
Area of Triangles	240
Parts of a Circle	241
Circumference	242
Area of a Circle	243
Area and Circumference Mixed Practice	244
Two-Step Area Problems	245
Measuring to Find Perimeter and Area	247
Estimating Area	248
Similar and Congruent	249
Solving Proportions	250
Similar Triangles	251
Pythagorean Theorem	254
Chapter Review	256
Chapter Test	258

CHAPTER 18 — 262
Solid Geometry

Understanding Volume	262
Volume of Cubes	263
Volume of Rectangular Prisms	264
Volume of Spheres, Cones, Cylinders, and Pyramids	265
Estimating Volume	267
Surface Area	
Cubes and Rectangular Prisms	268
Pyramids	270
Cylinders	271
Solid Geometry Word Problems	272
Chapter Review	273
Chapter Test	274

CHAPTER 19 — 276
Graphing

Graphing on a Number Line	276
Plotting Points on a Vertical Number Line	280
Cartesian Coordinates	281
Identifying Ordered Pairs	282
Drawing Geometric Figures on a Cartesian Coordinate Plane	284
Chapter Review	287
Chapter Test	288

CHAPTER 20 — 290
Transformations & Symmetry

Reflections	290
Rotations	293
Translations	295
Transformation Practice	297
Symmetry	298
Symmetry Practice	300
Chapter Review	301
Chapter Test	302

POST TEST — 304

Preface

Mathematics Review is one of several books in the ***Basics Made Easy*** series. This book will help students who need additional practice to achieve mastery of arithmetic, data interpretation, problem solving, and introductory algebra and geometry skills. The materials in this book are based on the objectives covered on the *Iowa Test of Basic Skills* (Levels 12, 13, & 14) and the *Stanford 9 Achievement Test* (Level Intermediate 2 through Advanced 2). In addition, ***Mathematics Review*** teaches the concepts and skills emphasized on end-of-grade tests and on high school exit exams. Each chapter contains concise lessons and frequent practice exercises. Each chapter ends with an open-ended chapter review followed by a multiple-choice chapter test in a standardized test format.

We welcome comments and suggestions about the book. Please contact the author at:

American Book Company
PO Box 2638
Woodstock, GA 30188-1383

Toll Free: 1 (888) 264-5877
Phone (770) 928-2834
Fax (770) 928-7483
Web site: www.americanbookcompany.com

ABOUT THE AUTHOR

Colleen Pintozzi has taught mathematics at the middle school, junior high, senior high and adult level for 22 years. She holds a B.S. degree from Wright State University in Dayton, Ohio and has done graduate work at Wright State, Duke University and the University of North Carolina at Chapel Hill. She is the author of **Passing the Georgia High School Graduation Test in Mathematics; Passing the Florida High School Competency Test; Passing the South Carolina Exit Exam in Mathematics; Passing the Alabama Graduation Examination in Mathematics; California Mathematics Review; Tackling the TAKS 11 Mathematics, parts 1 and 2; Passing the ISTEP+ Graduation Qualifying Examination in Mathematics; Ohio Graduation Test: Mathematics Review; Passing the Leap 21 Graduation Exit Examination in Mathematics; Passing the Nevada High School Proficiency Test in Mathematics; Passing the Minnesota Basic Standards Test in Mathematics; and Passing the TCAP Competency Test in Mathematics.**

DIAGNOSTIC EXAM FOR
BASICS MADE EASY MATHEMATICS

The Diagnostic Exam for Basics Made Easy Mathematics consists of 50 multiple-choice questions. You will be given as much time as you need for the test. You may refer to the formula chart on the following page as often as needed.

FORMULA SHEET

Formulas that you may need to answer questions on this exam are found below.

Area of a square = s^2

Area of a rectangle = lw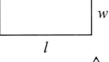

Area of a triangle = $\frac{1}{2}bh$

Area of a circle = πr^2

Circumference = πd or $2\pi r$

π = Pi = 3.14 or $\frac{22}{7}$

Area of a trapezoid = $\frac{(b_1 + b_2) \times h}{2}$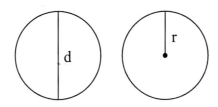

Volume of a rectangular prism = $l \times w \times h$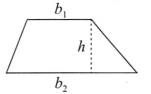

Volume of a cylinder = $\pi r^2 h$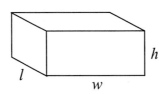

Pythagorean Theorem: $a^2 + b^2 = c^2$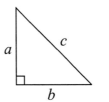

Diagnostic Exam For Basics Made Easy Mathematics

1. In the Venn diagram above, how many members are in the following set?

   {sister ∪ pet dog}

   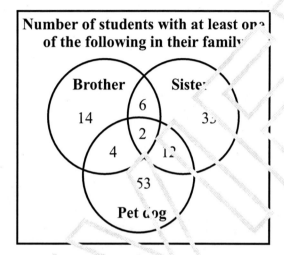

   A. 12
   B. 14
   C. 100
   D. 110

2. Which of the sets below does not contain {a, e, o} as a subset?

   A. {letters of the alphabet}
   B. {vowels in the alphabet}
   C. {letters in the word "predator"}
   D. {the first 5 letters of the alphabet}

3. Which of the following statements is false?

   A. {t, o} ⊂ {letters in the word "today"}
   B. 8 ∉ {0, 2, 4, 6, 8}
   C. ∅ = { }
   D. {a, b, c} ∩ {b, c, d, e} = {b, c}

4. Which of the following lists **all** the factors of the number 12?

   A. 1, 12
   B. 2, 3, 4, 6
   C. 5, 11
   D. none of the above

5. Mrs. Campbell's 5th grade class is going on a field trip. There are 29 children in the class. Parents are driving, and there will be 4 students per car. What is the smallest number of cars they will need for the children?

   A. 6
   B. 7
   C. 8
   D. 9

6. Ken had $90\frac{1}{2}$ feet of rope in his garage. He needed $73\frac{1}{3}$ feet to replace a rotted pulley rope. How many feet of rope did Ken have left?

   A. $17\frac{1}{5}$
   B. $17\frac{2}{5}$
   C. $18\frac{1}{6}$
   D. $17\frac{1}{6}$

7. Holly bought 10 yards of fabric to recover 6 dining room chairs. Each chair took $1\frac{1}{4}$ yards. How much fabric did she have left?

   A. $2\frac{1}{2}$
   B. 4
   C. $4\frac{4}{5}$
   D. $7\frac{1}{2}$

8. Which fraction is the smallest?

   A. $\frac{1}{2}$
   B. $\frac{2}{9}$
   C. $\frac{4}{7}$
   D. $\frac{1}{3}$

9. Hanna bought 3 pairs of socks priced at 3 for $5.00 and shoes for $45.95. She paid $2.55 sales tax. How much change did she receive from $100.00?

   A. $36.50
   B. $46.50
   C. $51.50
   D. $53.50

10. $\frac{3}{8}$ written as a decimal is

    A. 0.375
    B. 0.38
    C. 0.0375
    D. 0.3

11. Rami has an aquarium with 3 black goldfish and 4 orange goldfish. He purchased 2 more black goldfish to add to his aquarium. What is the new ratio of black goldfish to total goldfish?

    A. $\frac{2}{9}$
    B. $\frac{5}{4}$
    C. $\frac{4}{5}$
    D. $\frac{5}{9}$

12. Aunt Bess uses 3 cups of oatmeal to bake 6 dozen oatmeal cookies. How many cups of oatmeal would she need to bake 15 dozen cookies?

    A. 1.2
    B. 7.5
    C. 18
    D. 30

13. The spinner below stopped on the number 5 on the first spin. What is the probability that it will not stop on the number 5 on the second spin?

    A. $\frac{1}{5}$
    B. $\frac{1}{3}$
    C. $\frac{1}{6}$
    D. $\frac{5}{6}$

14. The shaded area of the graph below represents what percent of the total?

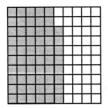

   A. 42%
   B. 58%
   C. 60%
   D. 580%

15. $\frac{3}{8}$ written as a percent is

   A. 37.5%
   B. 38%
   C. 3.75%
   D. 30%

16. John has been working at Palazone's Pizza for 6 months. He works 2 afternoons per week and all day Saturday and Sunday for a total of 24 hours per week. He has been promised a 6% raise. How much will he make per week now?

   A. $ 48.00
   B. $144.00
   C. $288.00
   D. Not enough information is given.

17. Elaine sells bread she bakes at home in three local food stores. She figured it is costing her 36¢ per loaf to make the bread, and she sells it for $1.25 per loaf. She works 25 hours a week.

   To find out how much Elaine makes an hour, you also need to know

   A. how many days she works in a week.
   B. how many days a loaf of bread can stay on the shelf before the expiration date.
   C. how many loaves she sells in a week.
   D. how far away the stores are from her home.

18. Renada had 2 twenty dollar bills, 4 ten dollar bills, 5 five dollar bills, and 11 one dollar bills. She used seven bills to pay a beautician $52 for a permanent. How many ten dollar bills did she use?

    A. 1
    B. 2
    C. 3
    D. 4

19. What is 7,003,780 expressed in scientific notation?

    A. 7.00378×10^{5}
    B. 7.00378×10^{-5}
    C. 7.00378×10^{6}
    D. 7.00378×10^{-6}

20. Which of the following is equal to 30 millimeters?

    A. 0.03 meters
    B. 0.003 kilometers
    C. 300 centimeters
    D. 30,000 meters

21.   8 yd 2 ft 9 in
    + 1 yd 2 ft 10 in

    A. 9 yd 1 ft 7 in
    B. 10 yd 1 ft 7 in
    C. 10 yd 2 ft 7 in
    D. 11 yd 7 in

22.   3 hr 25 min 34 sec
    + 1 hr 34 min 30 sec

    A. 4 hr 59 min 4 sec
    B. 5 hr   4 sec
    C. 5 hr 1 min 4 sec
    D. 6 hr   4 sec

23.   4 minutes 22 seconds
    − 2 minutes 45 seconds

    A. 1 minute 37 seconds
    B. 1 minute 23 seconds
    C. 2 minutes 17 seconds
    D. 2 minutes 23 seconds

24. What would be the total cost of 3 pairs of Knee Hi's, 6 pairs of Anklets, and 3 pairs of Support Socks?

SOCKS-A-PLENTY WAREHOUSE		
Item	3 pair price	6 pair price
Anklets	$4.50	$8.50
Sport Socks	$6.50	$11.50
Support Socks	$9.00	$15.75
Knee Hi	$6.00	$11.25
Please add 10% for shipping and handling		

    A. $23.50
    B. $25.85
    C. $33.50
    D. $77.55

25. According to the mileage chart above, how far is Tallahassee from Orlando?

	Jacksonville	Miami	Orlando	Tallahassee	Tampa	West Palm Beach
Jacksonville	0	341	142	163	202	277
Miami	341	0	229	478	273	66
Orlando	142	229	0	257	84	165
Tallahassee	163	478	257	0	275	414
Tampa	202	273	84	275	0	226
West Palm Beach	277	66	165	414	226	0

    A. 84 miles
    B. 142 miles
    C. 257 miles
    D. 478 miles

26. Rachel kept track of how many scoops she sold of the five most popular flavors in her ice cream shop.

Flavor	Scoops
Vanilla Bean	30
Chunky Chocolate	36
Strawberry Coconut	44
Chocolate Peanut Butter	46
Mint Chocolate Chip	28

    Which flavor of ice cream is closest to the mean number?

    A. Vanilla Bean
    B. Chocolate Peanut Butter
    C. Chunky Chocolate
    D. Strawberry Coconut

27. Which of the following sets of numbers has a median of 42?

    A. { 60, 42, 37, 22, 19 }
    B. { 16, 28, 42, 48 }
    C. { 42, 64, 20 }
    D. { 12, 42, 40, 50 }

28. Find the mode of 7, 12, 15, 7, 4, 1, and 10.

    A. 7
    B. 8
    C. 1
    D. 6

29. $|-8| - |4| =$

    A. 4
    B. −4
    C. 12
    D. −12

30. Which of the following shows the <u>Inverse Property of Multiplication</u>?

    A. $4 \times 1 = 4$

    B. $4 \times (2 \times 1) = (4 \times 2) \times 1$

    C. $4 \times 2 = 2 \times 4$

    D. $4 \times \frac{1}{4} = 1$

31. Which expression means the same as this phrase?

    8 less than 4 times a number

    A. $8 - x = 4$
    B. $8 + 4x$
    C. $8(x + 4)$
    D. $4x - 8$

32. If $c = 5$ and $d = 3$, evaluate

    $c^2 - 3d$

    A. 1
    B. 14
    C. 16
    D. 19

33. Choose the phrase that means the same as: $5a - 2$

    A. two less than a increased by five
    B. the product of five and a plus two
    C. two less than the difference of 5 and a
    D. two less than the product of 5 and a

34. What would replace n in this number sentence to make the sentence true?

    $2n - 8 = 16$

    A. 4
    B. 6
    C. 8
    D. 12

35.

Which inequality is represented on the number line above?

A. $x > 6$
B. $x \geq 6$
C. $x < 6$
D. $x \leq 6$

36. The function rule is $3x(x+5)$

x	$f(x)$
-2	

A. -18
B. 18
C. -42
D. 42

37. Adrianna has 4 hats, 8 shirts, and 9 pairs of pants. Choosing one of each, how many different clothes combinations can she make?

A. 17
B. 32
C. 96
D. 288

38. How many different ways can Mrs. Smith choose 2 students to go to the library if she has 20 students in her class?

A. 40
B. 190
C. 380
D. 3,800

39. What is the measure of ∠ARC?

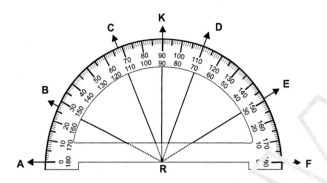

- A. 68°
- B. 72°
- C. 112°
- D. 128°

40. Which of the two angles are adjacent?

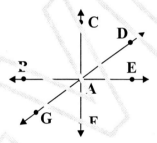

- A. ∠GAB and ∠DAE
- B. ∠CAG and ∠CAD
- C. ∠EAF and ∠CAP
- D. ∠CAE and ∠DAG

41. Which line segments are perpendicular in the figure below?

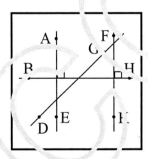

- A. $\overline{AE}$ and $\overline{BH}$
- B. $\overline{GD}$ and $\overline{FK}$
- C. $\overline{AE}$ and $\overline{GD}$
- D. $\overline{BH}$ and $\overline{GD}$

42. On the way to school, Bobby noticed the school bus travels 8 miles north, then turns west and goes 6 miles to get to school. If there were a road that went straight from Bobby's house to school, how long would the road be?

    A. 4 miles
    B. 9 miles
    C. 10 miles
    D. 14 miles

43.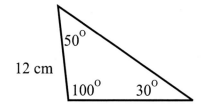

    Which triangle is **similar** to the triangle above?

    A.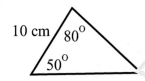

    B.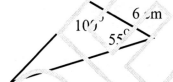

    C.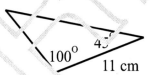

    D.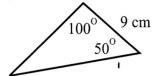

44. What is the volume of a cube that is 7 inches on each edge?

   A. 42 in³
   B. 49 in³
   C. 84 in³
   D. 343 in³

45. A rectangular box and a rectangular pyramid have the same dimensions for their bases and heights. How does the volume of the box compare to the volume of the pyramid?

   A. The volumes are the same.
   B. The volume of the box is twice the volume of the pyramid.
   C. The volume of the box is three times as large as the pyramid.
   D. The volume of the box is four times as large as the pyramid.

46.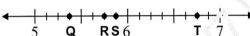

   Which point on the number line above represents $5\frac{3}{4}$?

   A. Point Q
   B. Point R
   C. Point S
   D. Point T

47.

   Where is $-2\frac{3}{4}$ located on the number line above?

   A. Point R
   B. Point S
   C. Point T
   D. Point U

48.

Randy plotted points G and H on the Cartesian coordinate graph above. Where could he plot points J and K if he wants to form a square GHJK?

A. J at (2, 4) and K at (−1, 4)
B. J at (3, 0) and K at (0, 0)
C. J at (2, 0) and K at (−1, 0)
D. J at (2, −1) and K at (−1, −1)

49. What kind(s) of symmetry does the following figure have? Choose the best answer.

I. reflectional symmetry
II. $\frac{1}{4}$ rotational symmetry
III. translational symmetry

A. I
B. II
C. I and II
D. III

50. What kind(s) of symmetry does the following figure have? Choose the best answer.

   A. ½ rotational symmetry
   B. reflectional symmetry
   C. translational symmetry
   D. no symmetry

EVALUATION CHART
DIAGNOSTIC TEST FOR BASICS MADE EASY MATHEMATICS

Directions: On the following chart, circle the question numbers that you answered incorrectly, and evaluate the results. Then turn to the appropriate topics (listed by chapters), read the explanations, and complete the exercises. Review the other chapters as needed.

		QUESTIONS	PAGES
Chapter 1:	Sets	1, 2, 3	19-24
Chapter 2:	Whole Numbers	4, 5	25-40
Chapter 3:	Fractions	6, 7, 8	41-58
Chapter 4:	Decimals	9, 10	59-81
Chapter 5:	Ratios, Probability, Proportions, And Scale Drawings	11, 12, 13	82-92
Chapter 6:	Percents	14, 15	93-108
Chapter 7:	Problem-Solving and Critical Thinking	16, 17, 18	109-126
Chapter 8:	Using Exponents and The Metric System	19, 20	127-137
Chapter 9:	Customary Measurements	21, 22, 23	138-146
Chapter 10:	Data Interpretation	24, 25	147-162
Chapter 11:	Statistics	26, 27, 28	163-174
Chapter 12:	Integers and Order of Operations	29, 30	175-185
Chapter 13:	Introduction to Algebra	31, 32, 33	186-196
Chapter 14:	Algebra	34, 35	197-210
Chapter 15:	Functions, Number Patterns, Permutations, and Combinations	36, 37, 38	211-227
Chapter 16:	Angles	39, 40	228-236
Chapter 17:	Plane Geometry	41, 42, 43	237-261
Chapter 18:	Solid Geometry	44, 45	262-275
Chapter 19:	Graphing	46, 47, 48	276-289
Chapter 20:	Transformations and Symmetry	49, 50	290-303

NOTES

Chapter 1

SETS

A set contains an object or objects that are clearly identified. The objects in a set are called **elements** or **members**. A set can also be **empty**. An **empty** set is also called the **null** set. The symbol ∅ or { } is used to denote the **null** set.

Members of a set can be described in two ways. One way is to give a rule of description that clearly defines the members. Another way is to make a roster of each member of the set, mentioning each member only <u>once</u> in any order. The members of a set are enclosed in braces { } and separated by commas.

                    Rule   **Roster**
     {the letters in the word "school"} = {s,c,h,o,l}

The symbol ∈ is used to show that an object is the member of a set. For example, $3 \in \{1,2,3,4\}$.
The symbol ∉ means "is **not** a member of". For example, $8 \notin \{1,2,3,4\}$.
The symbol ≠ means "is not equal to". For example,
          {the cities in Texas} ≠ {New York, Philadelphia, Nashville}

Read each of the statements below and tell whether they are true or false.

		True/False
1.	{the days of the week} = {January, March, December}	
2.	{The first five letters of the alphabet} = {a, b, c, d, e}	
3.	5 ∉ {all odd numbers}	
4.	Friday ∈ {the days of the week}	
5.	{the last three letters of the alphabet} ≠ {x, y, z}	
6.	t ∈ {the letters in the word "yellow"}	
7.	the letters in the word "funny" ∈ {the letters of the alphabet}	
8.	{The letters in the word "Alabama"} = {a, b, l, m}	
9.	{living unicorns} ≠ ∅	
10.	{the letters in the word "horse"} ≠ {the letters in the word "shore"}	

Identify each set by making a roster (see above). If a set has no members, use ∅.

11. {the letters in the word "hat" that are also in the word "thin"}
12. {the letters in the word "Mississippi"}
13. {the provinces of Canada that border the state of Texas}
14. {the letters in the word "kitchen" and also in the word "dinner"}
15. {the days of the week that have the letter "n"}
16. {the letters in the word "June" that are also in the word "April"}
17. {the letters in the word "instruments" that are not in the word "telescope"}
18. {the digits in the number "19,582" that are also in the number "56,871"}

SUBSETS

If every member of set *A* is also a member of set *B*, then set *A* is a **subset** of set *B*.

The symbol ⊂ means "is a subset of."

For example, every member of the set {1, 2, 3, 4} is a member of the set {1, 2, 3, 4, 5, 6, 7}. Therefore, {1, 2, 3, 4} ⊂ {1, 2, 3, 4, 5, 6, 7}.

The relationship can be pictured in the following diagram:

Set → 1 2 3 4
Subset → 5 6 7

The symbol ⊄ means "is **not** a subset of."

For example, **not** every member of the set {Ann, John, Sue} is a member of the set {Ann, John, Cindy}. Therefore, {Ann, John, Sue} ⊄ {Ann, John, Cindy}.

Read each of the statements below and tell whether they are true or false.

		True/False
1.	{a, b, c} ⊄ {a, b, c, d}	
2.	{1, 3, 5, 7} ⊂ {1, 2, 4, 5, 6, 7}	
3.	{a, e, i, o, u} ⊂ {the vowels of the alphabet}	
4.	{dogs} ⊂ {poodles, bull dogs, collies}	
5.	{fruit} ⊄ {apples, grapes, bananas, oranges}	
6.	{10, 20, 30, 40} ⊂ {20, 30, 40, 50}	
7.	{English, Spanish, French, German} ⊂ {languages}	
8.	{duck, swan, penguin} ⊄ {mammals}	
9.	{1, 2, 5, 10} ⊂ {whole numbers}	
10.	{Atlanta} ⊄ {U.S. state capitals}	

The set {Pam} has two subsets: {Pam} and ∅. The set {Emily, Brad} has 4 subsets: {Emily}, {Brad}, {Emily, Brad} and ∅. The ∅ is a subset of any set.

For each of the following sets, list all of the possible subsets.

1. {a}

2. {1, 2}

3. {r, s, t}

4. {1, 3, 5, 7}

5. {Joe, Ed}

INTERSECTION OF SETS

To find the intersection of two sets, you need to identify the members that the two sets share in common. The symbol for intersection is ∩. A **Venn diagram** shows how sets intersect.

Roster

{2, 4, 6, 8} ∩ {4, 6, 8, 10} = {4, 6, 8}

The shaded area is the intersection of the two sets. It shows which numbers both sets have in common.

Venn Diagram

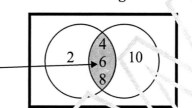

Find the intersection of the following sets:

1. {Ben, Jan, Dan, Tom} ∩ {Dan, Mike, Kate, Jan} =
2. {pink, purple, yellow} ∩ {purple, green, blue} =
3. {2, 4, 6, 8, 10} ∩ {1, 2, 3, 4, 5, 6, 7, 8, 9, 10} =
4. {a, e, i, o, u} ∩ {a, b, c, d, e, f, g, h, i, j, k} =
5. {pine, oak, walnut, maple} ∩ {maple, poplar} =
6. {100, 98, 95, 78, 62} ∩ {57, 62, 95, 98, 59} =
7. {orange, kiwi, coconut, pineapple} ∩ {pear, apple, orange} =

Look at the Venn diagram at the right to answer the questions below. Show your answers in roster form. Do the problem in parentheses first.

8. A ∩ B =
9. (A ∩ B) ∩ C =
10. A ∩ C =
11. B ∩ C =
12. (A ∩ C) ∩ B =

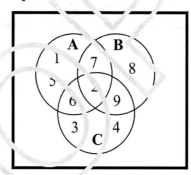

13. blue ∩ green =
14. (purple ∩ blue) ∩ green =
15. blue ∩ purple =
16. (blue ∩ green) ∩ purple =
17. green ∩ purple =

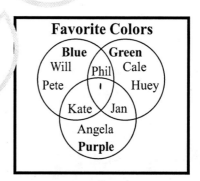

UNION OF SETS

The union of two sets means to put the members of two sets together into one set without repeating any members. The symbol for union is ∪.

$$\{1, 2, 3, 4\} \cup \{3, 4, 5, 6\} = \{1, 2, 3, 4, 5, 6\}$$

The union of theses two sets is the shaded area in the Venn diagram below.

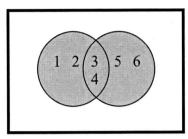

Find the union of the sets below.

1. {apples, pears, oranges} ∪ {pears, bananas, apples} =

2. {5, 10, 15, 20, 25} ∪ {10, 20, 30, 40} =

3. {Ted, Steve, Kevin, Michael} ∪ { Kevin, George, Kenny} =

4. {raisins, prunes, apricots} ∪ {peanuts, almonds, coconut} =

5. {sales, marketing, accounting} ∪ {receiving, shipping, sales} =

6. {beef, pork, chicken} ∪ {chicken, tuna, shark} =

Refer to the following Venn diagrams to answer the questions below. Identify each of the following sets by roster.

7. A ∪ C =

8. C ∪ B =

9. B ∪ A =

10. A ∪ B ∪ C =

11. North ∪ East =

12. North ∪ South =

13. East ∪ South =

14. North ∪ East ∪ South =

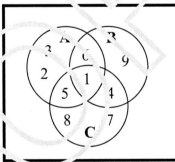

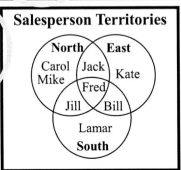

CHAPTER 18 REVIEW

Read each of the statements below and tell whether they are true of false.

		True or False
1.	{odd whole numbers} ⊂ {all whole numbers}	
2.	{yearly seasons} ≠ {spring, summer, fall, winter}	
3.	{Monday, Tuesday, Wednesday} ⊂ {days of the week}	
4.	United States of America ∉ {countries in North America}	
5.	{green, purple} ⊄ {primary colors}	
6.	{plants with red flowers} = ∅	
7.	{letters in "subsets"} = {b, u, s, e, t}	
8.	Milky Way ∉ {galaxies in the universe}	
9.	{Houston, Dallas} ⊂ {cities in Texas}	
10.	George Washington ∈ {former presidents of the United States}	

Complete the following statements.

11. {3, 6, 9, 12, 15} ∩ {0, 5, 10, 15, 20} =

12. {Felix, Mark, Kate} ∪ {Mark, Carol, Jack} =

13. {letters in "perfect"} ∩ {letters in "profit"} =

14. {Rome, London, Paris} ∩ {Italy, England, France} =

15. {black, white, gray} ∪ {red, white, blue} =

16. {1, 2, 3, 4, 5, 6} ∪ {2, 4, 6, 8, 10, 12} =

Refer to the Venn diagram to complete the following statements. Answers should be in roster form.

17. basketball ∪ football ∪ baseball =

18. basketball ∩ football =

19. (football ∩ basketball) ∩ baseball =

20. baseball ∪ basketball =

21. football ∪ baseball =

22. baseball ∩ football =

23. basketball ∩ baseball =

24. football ∪ basketball =

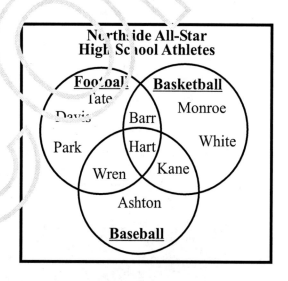

Northside All-Star High School Athletes

Use the following Venn diagram to answer questions 1 and 2.

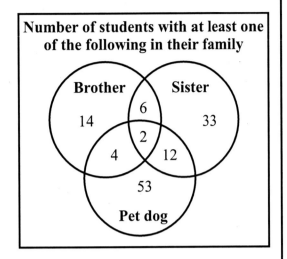

25. In the Venn diagram above, how many members are in the following set?
    {sister ∪ pet dog}

    A. 12
    B. 14
    C. 100
    D. 110

26. In the Venn diagram above, how many members are in the following set?
    {brother} ∩ {sister}

    A. 6
    B. 8
    C. 52
    D. 55

27. Which of the sets below does not contain {a, e, o} as a subset?

    A. {letters of the alphabet}
    B. {vowels in the alphabet}
    C. {letters in the word "predator"}
    D. {the first 5 letters of the alphabet}

28. Which of the following statements is true?

    A. {1, 2, 4, 5} ∪ ∅ = ∅
    B. {c} ∉ {c, a, t}
    C. {f, g} ⊂ {f, g, h, i}
    D. {k, l, m} ∪ ∅ = ∅

29. Which of the following statements is false?

    A. {t, o} ⊂ {letters in the word "today"}
    B. 8 ∉ {0, 2, 4, 6, 8}
    C. ∅ = { }
    D. {a, b, c} ∩ {b, c, d, e} = {b, c}

30.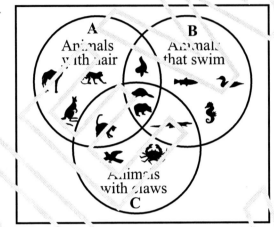

    According to the above diagram, which one of the following statements is false?

    A. $A \cap B$ = { 🐦, 🐢, 🐻 }
    B. 🐊 ∈ B
    C. { 🦘, 🐻, 🐱 } ⊂ A
    D. $B \cup C \neq \emptyset$

31. Which of the following sets equals ∅?

    A. {all whole numbers}
    B. {all letters in the alphabet}
    C. {all fish with feathers}
    D. {all flowering plants}

32. {Karen, John, Sue} ∩ {John, Perry, Kay}

    A. ∅
    B. {John}
    C. {Karen, Sue, Perry, Kay}
    D. {Karen, John, Sue, Perry, Kay}

Chapter 2

WHOLE NUMBERS

ADDING WHOLE NUMBERS

EXAMPLE: Find $302 + 54 + 712 + 9$

Step 1 Remember when you add to arrange the numbers in columns with the ones digits at the right.

$$\begin{array}{r} 302 \\ 54 \\ 712 \\ + 9 \end{array}$$

Step 2 Start at the right and add each column. Remember to carry when necessary.

$$\begin{array}{r} 1 \\ 302 \\ 54 \\ 712 \\ + 9 \\ \hline 1{,}077 \end{array}$$

Find the sum, and circle your answer.

1. $18 + 24 + 157$
2. $2{,}458 + 5{,}011$
3. $4{,}005 + 1{,}312$
4. $386 + 54 + 3$
5. $4{,}057 + 21 + 219$
6. $2{,}465 + 486$

7. The total of 9 and 104
8. 94 more than 541
9. 784 increased by 51
10. 18 more than 149
11. 5 more than 557
12. 102 added to 73

13. 298 increased by 23
14. 541 plus 102
15. $12 + 454 + 3 + 97$
16. The sum of 308 and 52
17. The total of 85, 78, and 215
18. $6 + 243 + 19$

SUBTRACTING WHOLE NUMBERS

EXAMPLE: Find 1006 − 568

Step 1 Remember when you subtract to arrange the numbers in columns with the ones digits at the right.

$$\begin{array}{r} 1006 \\ -568 \end{array}$$

Step 2 Start at the right, and subtract each column. Remember to borrow when necessary.

$$\begin{array}{r} {}^{9\,9}\!} \\ \cancel{100}^{1}\!6 \\ -\ 568 \\ \hline 438 \end{array}$$

← Borrow 1 from the 100, making it 99

Note: When you see "less than" in a problem, the second number becomes the top number when you set up the problem.

Find the difference, and circle your answer.

1. 541 − 35

2. 6007 − 279

3. 694 − 287

4. 902 − 471

5. 500 − 376

6. 1047 − 483

7. 14 less than 607

8. 881 decreased by 354

9. The difference between 284 and 29

10. 500 decreased by 125

11. 43 less than 752

12. 74 less than 1095

13. 96 less than 704

14. 327 less than 1002

15. The difference between 273 and 55

16. The difference between 2849 and 756

17. 975 decreased by 249

18. 405 decreased by 36

SUBTRACTING - BORROWING TWICE

EXAMPLE: Find 4034 − 365

Step 1 Arrange the numbers in columns with the ones digits at the right.

```
  4034
−  365
```

Step 2 Start at the right, and subtract each column. Remember to borrow when necessary.

$$4 0 \overset{2}{\cancel{3}} \overset{1}{4}$$
$$- \quad 3 6 5$$

Borrow 1 from 3. You cannot take 6 from 2 so...

Step 3

$$\overset{3}{\cancel{4}} \overset{9}{\cancel{0}} \overset{12}{\cancel{3}} \overset{1}{4}$$
$$- \quad 3 6 5$$
$$ 3 6 6 9$$

Borrow 1 from 40. Now finish subtracting.

Find the difference, and check by adding. When you add, the answer should be the same as the top number of the problem.

1. $\overset{9}{\cancel{1}}\overset{13}{\cancel{0}}46$
    -678
    $\overline{368}$ ⟩ add to check

2. 3186
    −395

3. 6416
    −3524

4. 7416
    − 846

5. 5417
    −583

6. 5442
    −679

7. 4379
    −889

8. 5462
    −4279

9. 3845
    −1174

10. 1724
    −576

11. 4043
    −2995

12. 2262
    −187

13. 3784
    −1285

14. 6175
    −462

15. 2547
    −1389

16. 1524
    −847

17. 9421
    −5841

18. 6703
    −505

19. 8674
    −4883

20. 7503
    −871

MULTIPLYING WHOLE NUMBERS

EXAMPLE: Multiply 256 × 73

Step 1 Line up the ones digits
Multiply 256 × 3.

$$\begin{array}{r} 1 \\ 256 \\ \times\ 7\boxed{3} \\ \hline 768 \end{array}$$

Step 2 Multiply 256 × 7.
Remember to shift the product one place to the left. Then add.

$$\begin{array}{r} 256 \\ \times\ \boxed{7}3 \\ \hline 768 \\ 1792 \\ \hline 18{,}688 \end{array}$$

Multiply:

1. 258
    × 72

2. 742
    × 44

3. 785
    × 32

4. 679
    × 35

5. 841
    × 27

6. 324
    × 19

7. 921
    × 22

8. 454
    × 56

9. 156
    × 95

10. 765
    × 94

11. 591
    × 25

12. 827
    × 56

13. 942
    × 24

14. 247
    × 84

15. 468
    × 43

16. 456
    × 47

17. 743
    × 65

18. 527
    × 38

19. 524
    × 39

20. 682
    × 64

DIVIDING WHOLE NUMBERS

EXAMPLE: Divide 4993 ÷ 24

Step 1 Rewrite the problem using the symbol ⟌.

Step 2 Divide 24 into 49. Multiply 2 × 24, and subtract.

Step 3 You will notice you **cannot** divide 24 into 19. You **must** put a 0 in the answer and then bring down the 3.

Step 4 Divide 24 into 193. Multiply 8 × 24, and subtract.

$$24\overline{)4993} \quad \begin{array}{r} 2 \\ -48 \\ \hline 19 \end{array}$$

$$24\overline{)4993} \quad \begin{array}{r} 20 \\ -48 \\ \hline 193 \end{array}$$

$$24\overline{)4993} \quad \begin{array}{r} 208\,r1 \\ -48 \\ \hline 193 \\ -192 \\ \hline 1 \end{array}$$

The answer is 208 with a remainder of 1.

Divide:

1. 6274 ÷ 13
2. 2384 ÷ 43
3. 12747 ÷ 24
4. 5417 ÷ 19

5. 8042 ÷ 27
6. 2548 ÷ 63
7. 6254 ÷ 41
8. 4362 ÷ 35

9. 4345 ÷ 25
10. 15467 ÷ 43
11. 7412 ÷ 54
12. 9379 ÷ 83

13. 9547 ÷ 31
14. 7436 ÷ 61
15. 5464 ÷ 38
16. 23567 ÷ 11

ROUNDING WHOLE NUMBERS

EXAMPLE:

Ten Thousands
Thousands
Hundreds
Tens
Ones

4 8,5 3 8

Consider the number 48,538 shown at the left with the place values labeled. To round to a given place value, first find the place in the number. Then look to the digit on the right. If the digit on the right is 5 or greater, INCREASE BY ONE the place to which the number is being rounded. All the digits to the right of the given place value become 0's. If the digit on the right is LESS THAN 5, leave that place value the same, and change the digits on the right to 0's.

EXAMPLE: Round the number 48,538 to the nearest:

ten	48,540
hundred	48,500
thousand	49,000
ten thousand	50,000

Round to the nearest ten.

1. 523 _____
2. 6,745 _____
3. 1,324 _____
4. 872 _____
5. 2,421 _____
6. 749 _____
7. 8,478 _____
8. 5498 _____

Round to the nearest hundred.

9. 659 _____
10. 4,478 _____
11. 32,197 _____
12. 12,651 _____
13. 847 _____
14. 3,045 _____
15. 752 _____
16. 45,327 _____
17. 15,602 _____
18. 9,485 _____
19. 41,974 _____
20. 3,649 _____

Round to the nearest thousand.

21. 54,985 _____
22. 62,743 _____
23. 11,342 _____
24. 7,153 _____
25. 82,571 _____
26. 132,250 _____
27. 75,209 _____
28. 20,093 _____
29. 68,987 _____
30. 822,621 _____
31. 42,105 _____
32. 4,902 _____

Round to the nearest ten thousand.

33. 28,235 _____
34. 39,078 _____
35. 242,469 _____
36. 153,958 _____
37. 458,347 _____
38. 50,512 _____
39. 85,369 _____
40. 471,963 _____

Round to the nearest dollar.

41. $18.52 _____
42. $16.23 _____
43. $114.99 _____
44. $18.13 _____

Round to the nearest ten dollars.

45. $19.56 _____
46. $48.63 _____
47. $64.25 _____
48. $84.95 _____

ESTIMATED SOLUTIONS

In the real world, estimates can be very useful. The best approach to finding estimates is to round off all numbers in the problem. Then solve the problem, and choose the closest answer. If money problems have both dollars and cents, round to the nearest dollar or ten dollars. $44.86 rounds to $40.

EXAMPLE: Which is a reasonable answer?   1580 ÷ 21

  A. 80   **B.** 800   **C.** 880   **D.** 8,000

Step 1 Round off the numbers in the problem.   1580 rounds to 1600   21 rounds to 20

Step 2 Work the problem.   1600 ÷ 20 = 80   The closest answer is **A.** 80.

Choose the best answer below.

1. Which is a reasonable answer?   544 × 12   **A.** 54   **B.** 500   **C.** 540   **D.** 5400

2. Jeff bought a pair of pants for $45.95, a belt for $12.97 and a dress shirt for $24.87. Estimate about how much he spent.   **A.** $60   **B.** $70   **C.** $80   **D.** $100

3. For lunch, Marcia ate a sandwich with 187 calories, a glass of skim milk with 121 calories, and 2 brownies with 102 calories each. About how many calories did she consume?
   A. 300   **B.** 350   **C.** 480   **D.** 510

4. Which is a reasonable answer?   89,990 ÷ 28
   A. 300   **B.** 500   **C.** 1,000   **D.** 3,000

5. Which is a reasonable answer?   74,295 − 62,304
   A. 12,000   **B.** 11,000   **C.** 10,000   **D.** 1,000

6. Delia bought 4 cans of soup at 99¢ each, a box of cereal for $4.78, and 2 frozen dinners at $3.89 each. About how much did she spend?
   A. $10.00   **B.** $11.00   **C.** $13.00   **D.** $17.00

7. Which is the best estimate?   22,480 + 5,516
   A. 2,800   **B.** 17,000   **C.** 28,000   **D.** 32,000

School Store Price List

Pencils	Erasers	Folders	Binders	Compass	Protractor	Paper	Pens
2 for 78¢	59¢	21¢	$2.79	$1.59	89¢	$1.29	$1.10

8. Jake needs 2 pencils, 3 erasers, a binder, and a compass. About how much money will he need according to the chart?   **A.** $7.00   **B.** $8.00   **C.** $9.00   **D.** $10.00

9. Tracy needs a pack of paper, 2 folders, a protractor, and 6 pencils. About how much money does she need?   **A.** $3.00   **B.** $4.00   **C.** $5.00   **D.** $6.00

10. Which is the best estimate?   23895 ÷ 599   **A.** 4   **B.** 40   **C.** 400   **D.** 4,000

PRIME AND COMPOSITE NUMBERS

A **prime** number is a number greater than 1 that can only be divided by itself and 1 without a remainder. For example, 17 is a **prime** number. 17 can only be divided by 1 and 17. 1 and 17 are the **factors** of 17.

A **composite** number is a number greater than 1 that can be divided by 1, itself, and at least one other number. 16 is a **composite** number. 16 can be divided by 1,2,4,8, and 16. 1,2,4,8, and 16 are the **factors** of 16.

List the **factors** of each number below. Then write **P** if it is a **prime** number or **C** if it is a **composite** number.

	Number	Factors	Prime or Composite
1.	24	1, 2, 3, 4, 6, 8, 12, 24	C
2.	33		
3.	25		
4.	42		
5.	14		
6.	35		
7.	47		
8.	49		
9.	18		
10.	56		
11.	71		
12.	52		
13.	19		
14.	31		
15.	26		
16.	16		
17.	21		
18.	41		
19.	90		
20.	45		
21.	13		
22.	9		
23.	12		
24.	27		

TIME OF TRAVEL

EXAMPLE: Katrina drove 384 miles at an average of 64 miles per hour. How many hours did she travel?

Solution: Divide the number of miles by the miles per hour. $\dfrac{384 \text{ miles}}{64 \text{ miles/hour}} = 6 \text{ hours}$

Find the hours of travel in each problem below.

1. Bobbi drove 342 miles at an average speed of 57 miles per hour. How many hours did she drive? _____

2. Jan set her speed control at 55 miles per hour and drove for 165 miles. How many hours did she drive? _____

3. John traveled 2,092 miles in a jet that flew an average of 523 miles per hour. How long was he in the air? _____

4. How long will it take a bus averaging 54 miles per hour to travel 378 miles? _____

5. Kyle drove his motorcycle in a 225 mile race, and he averaged 75 miles per hour. How long did it take for him to complete the race? _____

6. Stacy drove 576 miles at an average speed of 48 miles an hour. How many hours did she drive? _____

7. Kendra flew 250 miles in a glider and averaged 125 miles per hour in speed. How many hours did she fly? _____

8. Travis traveled 496 miles at an average speed of 62 miles per hour. How long did he travel? _____

9. Wanda rode her bicycle an average of 15 miles an hour for 60 miles. How many hours did she ride? _____

10. Rami drove 184 miles at an average speed of 46 miles per hour. How many hours did he drive? _____

11. A train traveled at a constant 85 miles per hour for 425 miles. How many hours did the train travel? _____

12. How long was Amy on the road if she drove 195 miles at an average of 65 miles per hour? _____

RATE

EXAMPLE: Laurie traveled 312 miles in 6 hours. What was her average rate of speed?

Solution: Divide the number of miles by the number of hours. $\frac{312 \text{ miles}}{6 \text{ hours}} = 52$ miles/hour

Laurie's average rate of speed was 52 miles per hour (or 52 mph).

Find the average rate of speed in each problem below.

1. A race car went 500 miles in 4 hours. What was its average rate of speed?

2. Carrie drove 124 miles in 2 hours. What was her average speed?

3. After 7 hours of driving, Ed had gone 364 miles. What was his average speed?

4. Vera drove 360 miles in 8 hours. What was her average speed?

5. After 3 hours of driving, Paul had gone 183 miles. What was his average speed?

6. Sarah ran 25 miles in 5 hours. What was her average speed?

7. A train traveled 492 miles in 6 hours. What was its average rate of speed?

8. A commercial jet traveled 1,572 miles in 3 hours. What was its average speed?

9. Vanessa drove 195 miles in 3 hours. What was her average speed?

10. Greg drove 8 hours from his home to a city 336 miles away. At what average speed did he travel?

11. Michael drove 64 miles in one hour. What was his average speed in miles per hour?

12. After 9 hours of driving, Kate had traveled 405 miles. What speed did she average?

DISTANCE

EXAMPLE: Jessie traveled for 7 hours at an average rate of 58 miles per hour. How far did she travel?

Solution: Multiply the number of hours by the average rate of speed.
7 hours × 58 miles/hour = 406 miles

Find the distance in each of the following problems.

1. Myra traveled for 9 hours at an average rate of 45 miles per hour. How far did she travel?

2. A tour bus drove 4 hours averaging 58 miles per hour. How many miles did it travel?

3. Tina drove for 7 hours at an average speed of 53 miles per hour. How far did she travel?

4. Michael raced for 3 hours averaging 176 miles per hour. How many miles did he race?

5. Kris drove 5 hours and averaged 49 miles per hour. How far did she travel?

6. Oliver drove at an average of 53 miles per hour for 3 hours. How far did he travel?

7. A commercial airplane traveled 514 miles per hour for 2 hours. How far did it fly?

8. A train traveled at 125 miles per hour for 4 hours. How many miles did it travel?

9. Carmen drove a constant 65 miles an hour for 3 hours. How many miles did he drive?

10. Jasmine drove for 5 hours averaging 40 miles per hour. How many miles did she drive?

11. Roger flew his glider for 2 hours at 87 miles per hour. How many miles did his glider fly?

12. Beth traveled at a constant 65 miles per hour for 4 hours. How far did she travel?

MILES PER GALLON

EXAMPLE: The odometer on Ginger's car read 46,789 before she started her trip. At the end of her trip, it read 47,119. She used 10 gallons of gasoline. How many miles per gallon did she average?

Step 1: Subtract the ending odometer reading from the beginning odometer reading.
47,119 − 46,789 = 330 miles traveled

Step 2: Divide the number of miles traveled by the number of gallons of gasoline used.
$\frac{330 \text{ miles}}{10 \text{ gallons}}$ = 33 miles per gallon

---

Compute the number of miles per gallon in each of the following problems.

1. Rachael's odometer read 125,625 at the beginning of her trip. At the end of her trip, it read 125,863. She used 7 gallons of gasoline. How many miles per gallon did she average? _____

2. Tamera traveled 492 miles on 12 gallons of gasoline. What was her average gas mileage? _____

3. The odometer on Blake's car read 3,973 before she started her trip. At the end of her trip, it read 4,625. She used 26 gallons of gasoline. How many miles per gallon did she average? _____

4. Farmer Joe's tractor odometer read 218,754 before tilling his fields. After tilling, it read 218,802. He used 4 gallons of gasoline. How many miles per gallon did his tractor average? _____

5. When Devyn started the week, the odometer on his van read 64,742. At the end of the week, it read 64,984. He used 11 gallons of gasoline. How many miles per gallon did he average? _____

6. Kathryn drove 364 miles on 13 gallons of gasoline. What was her gas mileage? _____

7. Manny drove his jeep to the beach towing his boat. His odometer read 23,745 before his trip, and it read 24,030 once he arrived at the beach. His jeep used 15 gallons of gas. How many miles per gallon did he average? _____

8. Bonnie's odometer read 17,846 before she drove to visit her aunt. Once she arrived at her aunt's house, her odometer read 18,726. She used 22 gallons of gas. What was her average gas mileage? _____

9. Ron traveled 74 miles on 2 gallons of gas. How many miles per gallon did he average? _____

10. Before Janet and Bill left for their vacation, their odometer read 87,985. When they arrived back home, their odometer read 88,753. They used 24 gallons of gasoline. What was their average gas mileage? _____

WORD PROBLEMS

1. If Jacob averages 15 points per basketball game, how many points will he score in a season with 12 games?

2. A cashier can ring up 12 items per minute. How long will it take the cashier to ring up a customer with 72 items?

3. Mrs. Randolph has 26 students in 1st period, 32 students in 2nd period, 27 students in 3rd period, and 30 students in 4th period. What is the total number of students Mrs. Randolph teaches?

4. When Gerald started on his trip, his odometer read 109,875. At the end of his trip it read 115,480. How many miles did he travel?

5. The Beta Club is raising money by selling boxes of candy. It sold 152 boxes on Monday, 256 boxes on Tuesday, 107 boxes on Wednesday, and 93 boxes on Thursday. How many total boxes did the Beta Club sell?

6. Jonah won 1056 tickets in the arcade. He purchased a pair of binoculars for 964 tickets. How many tickets does he have left?

7. A school cafeteria has 52 tables. If each table seats 14 people, how many people can be seated in the cafeteria?

8. Leadville, Colorado is 14,286 feet above sea level. Denver, Colorado is 5,280 feet above sea level. What is the difference in elevation between these two cities?

9. The local bakery made 288 donuts on Friday morning. How many dozen donuts did they make?

10. Mattie ate 14 chocolate-covered raisins. Her big brother ate 5 times as many. How many chocolate-covered raisins did her brother eat?

11. Concession stand sales for a football game totaled $1563. The actually cost for the food and beverages was $395. How much profit did the concession stand make?

12. An orange grove worker can harvest 480 oranges per hour by hand. How many oranges can the worker harvest in an 8 hour day?

CHAPTER REVIEW

1. Add: $18 + 694 + 123 + 75$

2. Subtract: $943 - 768$

3. Multiply: 452×23

4. Divide: $786 \div 95$

Round to the nearest 100:

5. 215
6. 553
7. 345
8. 351
9. 929
10. 872

Round to the nearest 10:

11. 67
12. 951
13. 683
14. 42
15. 1259
16. 563

Round to the nearest dollar:

17. $13.65
18. $22.12
19. $45.97
20. $18.03
21. $20.54
22. $53.49

Round to the nearest ten dollars:

23. $42.97
24. $65.03
25. $184.99
26. $56.22
27. $31.95
28. $47.82

Label the following numbers as C for composite or P for prime. If composite, list all factors.

29. 15
30. 16
31. 19
32. 20
33. 21
34. 23
35. 25

36. Tisha bought a skirt for $24.99, a blouse for $15.97, and 2 pairs of hose for $1.49 each. About how much money did she spend on her entire purchase?

37. A textile manufacturer uses 170 ft^2 of fabric to make a set of sheets. How many sets of sheets can it make from a 3840 ft^2 roll of fabric?

38. The Ring family's odometer read 65453 before driving to Disney World for vacation. After their vacation, the odometer read 66245. How many miles did they drive during their vacation?

39. Jonathan can assemble 47 widgets per hour. How many can he assemble in an 8 hour day?

40. Jacob drove 252 miles, and his average speed was 42 miles per hour. How many hours did he drive?

41. The Jones family traveled 300 miles in 5 hours. What was their average speed?

42. At a constant speed of 55 miles per hour, Connie drove for 2 hours. How many total miles did she travel?

43. Julie drove 189 miles and used 9 gallons of gas. What was her gas mileage?

CHAPTER TEST

1. Elaina was selling donuts for a school fund raiser. She sold 66 donuts and had 32 left. How many donuts did she have to begin with?

    A. 32
    B. 34
    C. 66
    D. 98

2. 　　28,452
    − 19,763

    A. 1,689
    B. 8,689
    C. 11,311
    D. 8,621

3. 　　406
    × 109

    A. 5,014
    B. 7,694
    C. 44,254
    D. 47,344

4. $7\overline{)11473}$

    A. 163
    B. 639
    C. 1,639
    D. 1,640

5. The sum of 464 and 298

    A. 156
    B. 652
    C. 662
    D. 762

6. The difference between 832 and 145

    A. 687
    B. 697
    C. 787
    D. 797

7. 1242 ÷ 52

    A. 23 r 46
    B. 24
    C. 24 r 8
    D. 24 r 18

8. 2032 − 984

    A. 1048
    B. 1148
    C. 1058
    D. 1158

9. Round 5082 to the nearest hundred.

    A. 5000
    B. 5080
    C. 5100
    D. 6000

10. Round 751 to the nearest ten.

    A. 700
    B. 750
    C. 760
    D. 800

11. Which of the following numbers is prime?

    A. 9
    B. 11
    C. 12
    D. 15

12. Which of the following lists **all** the factors of the number 12?

    A. 1, 12
    B. 2, 3, 4, 6
    C. 5, 11
    D. none of the above

13. Which is a reasonable estimate of the answer?

    　　　　16,729 ÷ 100

    A. 16,700
    B. 1,670
    C. 167
    D. 17

14. Mrs. Campbell's 5th grade class is going on a field trip. There are 29 children in the class. Parents are driving, and there will be 4 students per car. What is the smallest number of cars they will need for the children?

    A. 6
    B. 7
    C. 8
    D. 9

15. Jed has 155 head of cattle. Each eats 31 pounds of silage every day. How much silage does Jed feed his cattle every day?

    A. 5 lb
    B. 3705 lb
    C. 4705 lb
    D. 4805 lb

16. Kenya is reading a novel. She read 25 pages on Monday, 32 pages on Tuesday, and 15 pages on Wednesday. How many total pages did she read?

    A. 57
    B. 62
    C. 72
    D. 47

17. Patty sold 24 out of a total of 100 boxes of cookies. How many does she have left?

    A. 76
    B. 80
    C. 124
    D. 240

18. Jackie drove 504 miles at an average of 63 miles per hour. How long did she drive?

    A. 7 hours
    B. 8 hours
    C. 9 hours
    D. 11 hours

19. Jerry set up 18 rows of chairs and put 9 chairs in each row. How many chairs did he set up?

    A. 27
    B. 81
    C. 162
    D. 189

20. Keva traveled 464 miles in 8 hours. What was her average rate of speed?

    A. 48 miles per hour
    B. 50 miles per hour
    C. 55 miles per hour
    D. 58 miles per hour

21. Daniel drove 4 hours at an average speed of 56 miles per hour. How many miles did he drive?

    A. 14 miles
    B. 60 miles
    C. 204 miles
    D. 224 miles

22. Eric's mom drove 11 miles each way to bring Eric to school in the morning and back home in the afternoon. How many miles did she drive for 10 days of school?

    A. 110 miles
    B. 220 miles
    C. 330 miles
    D. 440 miles

23. The Mickle family drove 252 miles on a tank of gas. If their gas tank holds 12 gallons of gas, what was their average gas mileage?

    A. 20 miles per hour
    B. 21 miles per hour
    C. 22 miles per hour
    D. 23 miles per hour

Chapter 3

FRACTIONS

SIMPLIFYING IMPROPER FRACTIONS

EXAMPLE 1: Simplify: $\frac{21}{4} = 21 \div 4 = 5$ remainder 1   The quotient, 5, becomes the whole number portion of the mixed number.

The remainder, 1, becomes the top number of the fraction.

$\frac{21}{4} = 5\frac{1}{4}$ The bottom number of the fraction always remains the same.

EXAMPLE 2: Simplify: $\frac{11}{6}$

Step 1 $\frac{11}{6}$ is the same as $11 \div 6$. $11 \div 6 = 1$ with a remainder of 5.

Step 2   Rewrite as a whole number with a fraction. $1\frac{5}{6}$

Simplify the following improper fractions.

1. $\frac{13}{5} =$ ___   4. $\frac{7}{6} =$ ___   7. $\frac{13}{8} =$ ___   10. $\frac{13}{4} =$ ___   13. $\frac{17}{9} =$ ___   16. $\frac{5}{2} =$ ___

2. $\frac{11}{3} =$ ___   5. $\frac{19}{6} =$ ___   8. $\frac{9}{5} =$ ___   11. $\frac{18}{7} =$ ___   14. $\frac{27}{8} =$ ___   17. $\frac{7}{4} =$ ___

3. $\frac{24}{6} =$ ___   6. $\frac{16}{7} =$ ___   9. $\frac{22}{3} =$ ___   12. $\frac{15}{2} =$ ___   15. $\frac{22}{7} =$ ___   18. $\frac{21}{10} =$ ___

Fractions that have the same denominator (bottom number) can be added quickly. Add the numerators (top numbers) and keep the bottom number the same. Simplify the answer. The first one is done for you.

19. $\frac{2}{9}$	20. $\frac{2}{6}$	21. $\frac{3}{8}$	22. $\frac{9}{10}$	23. $\frac{8}{13}$	24. $\frac{6}{7}$	25. $\frac{3}{5}$	26. $\frac{7}{12}$	27. $\frac{4}{11}$
$\frac{7}{9}$	$\frac{4}{6}$	$\frac{5}{8}$	$\frac{1}{10}$	$\frac{6}{13}$	$\frac{4}{7}$	$\frac{4}{5}$	$\frac{5}{12}$	$\frac{3}{11}$
$\frac{4}{9}$	$\frac{5}{6}$	$\frac{7}{8}$	$\frac{3}{10}$	$\frac{2}{13}$	$\frac{5}{7}$	$\frac{3}{5}$	$\frac{1}{12}$	$\frac{6}{11}$
$\frac{13}{9} = 1\frac{4}{9}$								

CHANGING MIXED NUMBERS TO IMPROPER FRACTIONS

EXAMPLE: Change $4\frac{3}{5}$ to an improper fraction.

Step 1 Multiply the whole number (4) by the bottom number of the fraction (5). $4 \times 5 = 20$

Step 2 Add the top number to the product from step 1. $20 + 3 = 23$

Step 3 Put the answer over the bottom number (5).

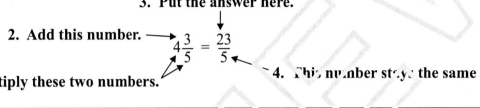

Change the following mixed numbers to improper fractions.

1. $3\frac{1}{2} =$ ___
2. $2\frac{7}{8} =$ ___
3. $9\frac{2}{3} =$ ___
4. $4\frac{3}{5} =$ ___

5. $7\frac{1}{4} =$ ___
6. $8\frac{5}{8} =$ ___
7. $5\frac{2}{7} =$ ___
8. $2\frac{4}{9} =$ ___

9. $6\frac{1}{5} =$ ___
10. $5\frac{2}{7} =$ ___
11. $3\frac{2}{5} =$ ___
12. $9\frac{3}{8} =$ ___

13. $10\frac{4}{5} =$ ___
14. $3\frac{3}{10} =$ ___
15. $4\frac{1}{7} =$ ___
16. $2\frac{5}{6} =$ ___

17. $7\frac{3}{7} =$ ___
18. $6\frac{7}{9} =$ ___
19. $7\frac{2}{5} =$ ___
20. $1\frac{6}{7} =$ ___

Whole numbers become improper fractions when you put them over 1. The first one is done for you.

21. $4 = \frac{4}{1}$
22. $10 =$ ___
23. $3 =$ ___
24. $2 =$ ___
25. $15 =$ ___
26. $5 =$ ___
27. $6 =$ ___
28. $11 =$ ___
29. $8 =$ ___
30. $16 =$ ___

GREATEST COMMON FACTOR

Find the greatest common factor (GCF) of 16 and 24.

To find the **greatest common factor (GCF)** of two numbers, first list the factors of each number.

The factors of 16 are: 1, 2, 4, 8, and 16.
The factors of 24 are: 1, 2, 3, 4, 6, 8, 12, and 24.

What is the **largest** number they both have in common? **8**
8 is the **greatest** (largest number) **common factor**.

Find all the factors and the greatest common factor (GCF) of each pair of numbers below.

	Pairs	Factors	GCF		Pairs	Factors	GCF
1.	10			10.	6		
	15				42		
2.	12			11.	14		
	16				63		
3.	18			12.	9		
	36				51		
4.	27			13.	18		
	45				45		
5.	32			14.	12		
	40				20		
6.	16			15.	16		
	48				40		
7.	14			16.	10		
	42				45		
8.	4			17.	18		
	26				30		
9.	8			18.	15		
	28				25		

REDUCING PROPER FRACTIONS

EXAMPLE: Reduce $\frac{4}{8}$ to lowest terms.

Step 1 First you need to find the greatest common factor of 4 and 8. Think: What is the largest number that can be divided into 4 and 8 without a remainder?

These must be the same number. $?\overline{)4}$ 4 and 8 can both be divided by 4.
$?\overline{)8}$

Step 2 Divide the top and bottom of the fraction by the same number.
$$\frac{4 \div 4}{8 \div 4} = \frac{1}{2}$$

Reduce the following fractions to lowest terms.

1. $\frac{2}{8} =$ _____
2. $\frac{12}{15} =$ _____
3. $\frac{9}{27} =$ _____
4. $\frac{12}{42} =$ _____
5. $\frac{3}{21} =$ _____
6. $\frac{27}{54} =$ _____

7. $\frac{14}{22} =$ _____
8. $\frac{9}{21} =$ _____
9. $\frac{4}{14} =$ _____
10. $\frac{6}{26} =$ _____
11. $\frac{30}{45} =$ _____
12. $\frac{16}{64} =$ _____

13. $\frac{10}{25} =$ _____
14. $\frac{3}{12} =$ _____
15. $\frac{15}{30} =$ _____
16. $\frac{12}{36} =$ _____
17. $\frac{13}{39} =$ _____
18. $\frac{28}{49} =$ _____

19. $\frac{8}{18} =$ _____
20. $\frac{14}{21} =$ _____
21. $\frac{2}{12} =$ _____
22. $\frac{5}{15} =$ _____
23. $\frac{9}{15} =$ _____
24. $\frac{24}{48} =$ _____

25. $\frac{3}{18} =$ _____
26. $\frac{6}{27} =$ _____
27. $\frac{4}{18} =$ _____
28. $\frac{2}{28} =$ _____
29. $\frac{14}{42} =$ _____
30. $\frac{18}{36} =$ _____

MULTIPLYING FRACTIONS

EXAMPLE: Multiply $4\dfrac{3}{8} \times \dfrac{8}{10}$

Step 1 Change the mixed numbers in the problem to improper fractions. $\dfrac{35}{8} \times \dfrac{8}{10}$

Step 2 When multiplying fractions, you can cancel and simplify terms that have a common factor.

The 8 in the first fraction will cancel with the 8 in the second fraction.

The terms 35 and 10 are both divisible by 5, so 35 simplifies to 7 and 10 simplifies to 2.

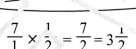

Step 3 Multiply the simplified fractions. $\dfrac{7}{1} \times \dfrac{1}{2} = \dfrac{7}{2} = 3\dfrac{1}{2}$

Step 4 You cannot leave an improper fraction as the answer. You must change it to a mixed number.

Multiply and reduce answers to lowest terms.

1. $3\dfrac{1}{5} \times 1\dfrac{1}{2}$

2. $\dfrac{3}{8} \times 3\dfrac{3}{7}$

3. $4\dfrac{1}{3} \times 2\dfrac{1}{4}$

4. $4\dfrac{2}{3} \times 3\dfrac{3}{4}$

5. $1\dfrac{1}{2} \times 1\dfrac{2}{5}$

6. $3\dfrac{3}{7} \times \dfrac{5}{6}$

7. $3 \times 6\dfrac{1}{3}$

8. $1\dfrac{1}{6} \times 8$

9. $6\dfrac{2}{5} \times 5$

10. $6 \times 1\dfrac{3}{8}$

11. $\dfrac{5}{7} \times 2\dfrac{1}{3}$

12. $1\dfrac{2}{5} \times 1\dfrac{1}{4}$

13. $2\dfrac{1}{2} \times 5\dfrac{4}{5}$

14. $7\dfrac{2}{3} \times \dfrac{3}{4}$

15. $2 \times 2\dfrac{1}{4}$

16. $3\dfrac{1}{8} \times 1\dfrac{3}{5}$

17. $2\dfrac{3}{4} \times 6\dfrac{2}{5}$

18. $5\dfrac{3}{5} \times 1\dfrac{1}{14}$

19. $2\dfrac{1}{3} \times 3\dfrac{3}{4}$

20. $5\dfrac{1}{9} \times 1\dfrac{1}{23}$

21. $6\dfrac{2}{5} \times 1\dfrac{9}{16}$

22. $2\dfrac{1}{7} \times 2\dfrac{1}{10}$

23. $3\dfrac{3}{10} \times 2\dfrac{2}{3}$

24. $\dfrac{3}{5} \times 4\dfrac{1}{6}$

DEDUCTIONS - FRACTION OFF

Sometimes sale prices are advertised as $\frac{1}{4}$ off or $\frac{1}{3}$ off. To find out how much you will save, just multiply the original price by the fraction off.

EXAMPLE: CD players are on sale for $\frac{1}{3}$ off. How much can you save on a $240 CD player?

$$\frac{1}{\underset{1}{\cancel{3}}} \times \frac{\overset{80}{\cancel{240}}}{1} = 80. \text{ You can save } \$80.00$$

Find the amount of savings in the problems below.

J.P. Nichols is having a liquidation sale on all furniture. Sale prices are $\frac{1}{2}$ off the regular price. How much can you save on the following furniture items?

Liquidation Furniture Sale

$\frac{1}{2}$ off all items in the store

Item	Regular Price	Savings
1. Couch	$850	_____
2. Loveseat	$624	_____
3. Recliner	$457	_____
4. Dining Room Set	$1352	_____
5. Bedroom Set	$2648	_____

Buy Rite Computer Store is having a $\frac{1}{3}$ off sale on selected computer items in the store. How much can you save on the following items?

Buy Rite Computer Sale

$\frac{1}{3}$ off selected items in the store

Item	Regular Price	Savings
6. Midline Computer	$1383	_____
7. Notebook Computer	$2220	_____
8. Tape Backup Drive	$210	_____
9. Laser Printer	$855	_____
10. Digital Camera	$690	_____

DIVIDING FRACTIONS

EXAMPLE: $1\frac{3}{4} \div 2\frac{5}{8}$

Step 1 Change the mixed numbers in the problem to improper fractions. $\frac{7}{4} \div \frac{21}{8}$

Step 2 Invert (turn upside down) the **second** fraction and multiply. $\frac{7}{4} \times \frac{8}{21}$

Step 3 Cancel where possible and multiply. $\frac{\cancel{7}^1}{\cancel{4}_1} \times \frac{\cancel{8}^2}{\cancel{21}_3} = \frac{2}{3}$

Divide and reduce answers to lowest terms.

1. $2\frac{2}{3} \div 1\frac{7}{9}$
2. $5 \div 1\frac{1}{2}$
3. $1\frac{5}{8} \div 2\frac{1}{4}$
4. $8\frac{2}{3} \div 2\frac{1}{6}$
5. $2\frac{4}{5} \div 2\frac{1}{5}$
6. $3\frac{2}{3} \div \frac{1}{6}$
7. $10 \div \frac{4}{5}$
8. $6\frac{1}{4} \div 1\frac{1}{2}$

9. $\frac{2}{5} \div 2$
10. $1\frac{1}{6} \div 1\frac{2}{3}$
11. $9 \div 2\frac{1}{4}$
12. $5\frac{1}{3} \div 2\frac{2}{5}$
13. $4\frac{1}{5} \div \frac{9}{10}$
14. $2\frac{2}{3} \div 4\frac{4}{5}$
15. $3\frac{3}{8} \div 3\frac{6}{7}$
16. $5\frac{1}{4} \div \frac{3}{4}$

17. $1\frac{2}{3} \div 1\frac{7}{8}$
18. $12 \div \frac{4}{7}$
19. $\frac{3}{5} \div \frac{4}{7}$
20. $4\frac{1}{3} \div 2\frac{8}{9}$
21. $5\frac{2}{5} \div 3\frac{1}{5}$
22. $1\frac{5}{6} \div \frac{1}{2}$
23. $10\frac{2}{3} \div 1\frac{7}{9}$
24. $4\frac{2}{7} \div 3$

FINDING NUMERATORS

REMEMBER: Any fraction that has the same non-zero Numerator (top number) and Denominator (bottom number) equals 1.

EXAMPLES: $\frac{5}{5}=1 \qquad \frac{8}{8}=1 \qquad \frac{12}{12}=1 \qquad \frac{15}{15}=1 \qquad \frac{25}{25}=1$

Any fraction multiplied by 1 in any fraction form remains equal.

EXAMPLES: $\frac{3}{7} \times \frac{4}{4} = \frac{12}{28}$ so $\frac{3}{7} = \frac{12}{28}$

PROBLEM: Find the missing numerator (top number). $\frac{5}{8} = \frac{}{24}$

Step 1 Ask yourself, "What was 8 multiplied by to get 24?" 3 is the answer.

Step 2 The only way to keep the fraction equal is to multiply the top and bottom number by the same number. The bottom number was multiplied by 3, so multiply the top number by 3. $\frac{5}{8} \times \frac{3}{3} = \frac{15}{24}$

Find the missing numerators from the following equivalent fractions.

1. $\frac{2}{6} = \frac{}{18}$
2. $\frac{2}{3} = \frac{}{27}$
3. $\frac{4}{9} = \frac{}{18}$
4. $\frac{7}{15} = \frac{}{45}$
5. $\frac{9}{10} = \frac{}{50}$

6. $\frac{5}{6} = \frac{}{36}$
7. $\frac{1}{4} = \frac{}{36}$
8. $\frac{?}{14} = \frac{}{28}$
9. $\frac{2}{5} = \frac{}{25}$
10. $\frac{4}{11} = \frac{}{33}$

11. $\frac{5}{6} = \frac{}{18}$
12. $\frac{6}{11} = \frac{}{22}$
13. $\frac{6}{15} = \frac{}{45}$
14. $\frac{1}{9} = \frac{}{18}$
15. $\frac{7}{8} = \frac{}{40}$

16. $\frac{1}{12} = \frac{}{48}$
17. $\frac{3}{8} = \frac{}{64}$
18. $\frac{3}{4} = \frac{}{16}$
19. $\frac{2}{7} = \frac{}{49}$
20. $\frac{11}{12} = \frac{}{24}$

21. $\frac{2}{5} = \frac{}{45}$
22. $\frac{4}{5} = \frac{}{15}$
23. $\frac{1}{9} = \frac{}{27}$
24. $\frac{3}{8} = \frac{}{56}$
25. $\frac{2}{13} = \frac{}{26}$

26. $\frac{1}{7} = \frac{}{35}$
27. $\frac{4}{5} = \frac{}{10}$
28. $\frac{3}{10} = \frac{}{40}$
29. $\frac{7}{8} = \frac{}{48}$
30. $\frac{6}{7} = \frac{}{14}$

ORDERING FRACTIONS

If the four fractions are given $\frac{3}{10}$, $\frac{5}{6}$, $\frac{1}{2}$, and $\frac{9}{10}$, the greatest, $\frac{9}{10}$, and least, $\frac{3}{10}$, are easy to find. If the four fractions given are $\frac{3}{5}$, $\frac{5}{6}$, $\frac{1}{2}$, and $\frac{11}{15}$, you probably need to find a common denominator first to figure out which fraction is the greatest or least.

$\frac{3}{5} = \frac{18}{30}$

$\frac{5}{6} = \frac{25}{30}$

$\frac{1}{2} = \frac{15}{30}$

$\frac{11}{15} = \frac{22}{30}$

Ask yourself what number you can divide by 5, 6, 2, and 15 without a remainder.

$5)\overline{}\ ?\quad 6)\overline{}\ ?\quad 2)\overline{}\ ?\quad 15)\overline{}\ ?$

The number 30 is the smallest number that can be divided by 5, 6, 2, and 15. Once equivalent fractions are found with the same denominator, you can look at the top number to tell which fraction is greatest $\frac{5}{6}$ or least, $\frac{1}{2}$.

Circle the greatest fraction and underline the least fraction in each group below.

1. $\frac{1}{4}$, $\frac{3}{8}$, $\frac{1}{12}$, $\frac{5}{6}$

2. $\frac{3}{5}$, $\frac{7}{10}$, $\frac{11}{20}$, $\frac{3}{4}$

3. $\frac{2}{3}$, $\frac{1}{6}$, $\frac{4}{9}$, $\frac{1}{2}$

4. $\frac{67}{100}$, $\frac{13}{25}$, $\frac{3}{10}$, $\frac{4}{5}$

5. $\frac{7}{9}$, $\frac{5}{12}$, $\frac{3}{4}$, $\frac{5}{8}$

6. $\frac{5}{18}$, $\frac{1}{4}$, $\frac{5}{9}$, $\frac{1}{2}$

7. $\frac{5}{6}$, $\frac{1}{2}$, $\frac{3}{4}$, $\frac{7}{12}$

8. $\frac{13}{36}$, $\frac{3}{4}$, $\frac{3}{13}$, $\frac{7}{9}$

9. $\frac{6}{7}$, $\frac{2}{3}$, $\frac{10}{21}$, $\frac{1}{2}$

10. $\frac{2}{15}$, $\frac{7}{8}$, $\frac{1}{4}$, $\frac{1}{2}$

Find common denominators to determine which fraction is between the two given fractions. Circle the answer.

11. Which fraction is between $\frac{1}{3}$ and $\frac{5}{8}$? $\frac{2}{3}$, $\frac{3}{4}$, $\frac{5}{6}$, or $\frac{3}{8}$

12. Which fraction is between $\frac{1}{2}$ and $\frac{5}{8}$? $\frac{1}{3}$, $\frac{1}{4}$, $\frac{3}{8}$, or $\frac{9}{16}$

13. Which fraction is between $\frac{3}{7}$ and $\frac{7}{8}$? $\frac{1}{4}$, $\frac{3}{4}$, $\frac{5}{14}$, or $\frac{25}{28}$

14. Which fraction is between $\frac{2}{3}$ and $\frac{8}{9}$? $\frac{1}{4}$, $\frac{5}{6}$, $\frac{5}{9}$, or $\frac{11}{18}$

LEAST COMMON MULTIPLE

Find the least common multiple (LCM) of 6 and 10.

To find the **least common multiple (LCM)** of two numbers, first list the multiples of each number. The multiples of a number are 1 times the number, 2 times the number, 3 times the number and so on.

The multiples of 6 are: 6, 12, 18, 24, 30 …

The multiples of 10 are: 10, 20, 30, 40, 50…

What is the smallest multiple they both have in common?   30
30 is the **least** (smallest number) **common multiple** of 6 and 10.

Find the least common multiple (LCM) of each pair of numbers below.

	Pairs	Multiples	LCM		Pairs	Multiples	LCM
1.	6	6, 12, 18, 24, 30	30	10.	6		
	15	15, 30			7		
2.	12			11.	4		
	16				18		
3.	18			12.	9		
	36				6		
4.	7			13.	30		
	3				45		
5.	12			14.	3		
	9				8		
6.	6			15.	12		
	8				9		
7.	2			16.	5		
	14				45		
8.	9			17.	3		
	6				5		
9.	2			18.	4		
	15				22		

ADDING FRACTIONS

EXAMPLE: Add $3\frac{1}{2} + 2\frac{2}{3}$

Step 1 Rewrite the problem vertically, and find a common denominator.

Think: What is the smallest number I can divide 2 and 3 into without a remainder? $2\overline{)?}$ $3\overline{)?}$ 6, of course.

$3\frac{1}{2} = \frac{}{6}$
$2\frac{2}{3} = \frac{}{6}$

Step 2 Find the missing numerators (top number) in the same way you did on page 48.

$3\frac{1}{2} = 3\frac{3}{6}$
$+2\frac{2}{3} = +2\frac{4}{6}$
$\phantom{+2\frac{2}{3}} = 5\frac{7}{6}$

Step 3 Add whole numbers and fractions and simplify.

$= 6\frac{1}{6}$

Add and simplify the answers.

1. $3\frac{5}{9} + 5\frac{2}{3}$
2. $1\frac{1}{4} + 4\frac{2}{5}$
3. $3\frac{3}{4} + 2\frac{3}{5}$
4. $2\frac{1}{4} + 1\frac{7}{8}$
5. $6\frac{5}{6} + 4\frac{1}{3}$
6. $9\frac{1}{5} + 5\frac{5}{6}$
7. $\frac{1}{3} + 7\frac{3}{4}$
8. $9\frac{4}{9} + 3\frac{2}{3}$
9. $4\frac{7}{10} + 8\frac{2}{3}$
10. $5\frac{2}{7} + \frac{1}{2}$
11. $3\frac{3}{11} + 2\frac{3}{4}$
12. $\frac{5}{5} + \frac{4}{9}$
13. $\frac{5}{8} + \frac{2}{3}$
14. $6\frac{4}{7} + 5\frac{1}{8}$
15. $9\frac{1}{3} + 4\frac{7}{8}$
16. $7\frac{3}{4} + 3\frac{5}{12}$
17. $\frac{1}{2} + \frac{2}{3}$
18. $6\frac{1}{2} + 2\frac{7}{8}$
19. $7\frac{2}{3} + 5\frac{3}{8}$
20. $3\frac{7}{8} + 9\frac{4}{5}$
21. $4\frac{7}{12} + \frac{2}{3}$

SUBTRACTING MIXED NUMBERS FROM WHOLE NUMBERS

EXAMPLE: Subtract $15 - 3\frac{3}{4}$

Step 1 Rewrite the problem vertically.
$$\begin{array}{r} 15 \\ -\ 3\frac{3}{4} \end{array}$$

Step 2 You cannot subtract three-fourths from nothing. You must borrow 1 from 15. You will need to put the 1 in fraction form. If you use $\frac{4}{4}$ ($\frac{4}{4} = 1$), you will be ready to subtract.

$$\begin{array}{r} 1\overset{4}{\cancel{5}}\frac{4}{4} \\ -\ 3\frac{3}{4} \\ \hline 11\frac{1}{4} \end{array}$$

Subtract

1. $12 - 3\frac{2}{9}$
2. $3 - 1\frac{4}{7}$
3. $24 - 11\frac{4}{5}$
4. $2 - 1\frac{2}{5}$
5. $4 - 1\frac{5}{8}$
6. $11 - 9\frac{7}{8}$
7. $14 - 9\frac{7}{12}$
8. $8 - 3\frac{1}{3}$
9. $5 - 3\frac{1}{2}$
10. $17 - 13\frac{1}{5}$
11. $3 - 1\frac{5}{11}$
12. $13 - 8\frac{9}{10}$
13. $15 - 6\frac{3}{4}$
14. $6 - 4\frac{8}{9}$
15. $20 - 12\frac{6}{7}$
16. $21 - 1\frac{3}{20}$
17. $9 - 5\frac{2}{3}$
18. $8 - 7\frac{3}{5}$
19. $5 - 4\frac{5}{8}$
20. $14 - 9\frac{1}{7}$
21. $12 - 4\frac{1}{6}$
22. $2 - 1\frac{2}{3}$
23. $42 - 30\frac{2}{9}$
24. $7 - 5\frac{9}{13}$
25. $9 - 1\frac{3}{8}$
26. $14 - 10\frac{5}{9}$
27. $16 - 8\frac{1}{4}$
28. $15 - 3\frac{5}{7}$

SUBTRACTING MIXED NUMBERS WITH BORROWING

EXAMPLE: Subtract $7\frac{1}{4} - 5\frac{5}{6}$

Step 1 Rewrite the problem and find a common denominator.

$$7\frac{1}{4} = \frac{3}{12}$$
$$-5\frac{5}{6} = \frac{10}{12}$$

Step 2 You cannot subtract 10 from 3. You must borrow 1 from the 7. The 1 will be in the fraction form $\frac{12}{12}$ which you must add to the $\frac{3}{12}$ you already have making $\frac{15}{12}$.

$$7\frac{1}{4} = \overset{6}{\cancel{7}}\frac{\overset{15}{\cancel{3}}}{12}$$
$$-5\frac{5}{6} = \frac{10}{12}$$
$$\overline{1\frac{5}{12}}$$

Step 3 Subtract whole numbers and fractions and simplify.

Subtract and simplify.

1. $4\frac{1}{3} - 1\frac{5}{9}$
2. $3\frac{1}{9} - 2\frac{5}{6}$
3. $8\frac{4}{7} - 5\frac{1}{3}$

4. $5\frac{2}{5} - 3\frac{1}{2}$
5. $8\frac{2}{5} - 5\frac{3}{10}$
6. $9\frac{2}{5} - 4\frac{3}{4}$

7. $9\frac{3}{4} - 2\frac{1}{3}$
8. $5\frac{1}{7} - \frac{2}{3}$
9. $6\frac{1}{5} - 3\frac{3}{8}$

10. $5\frac{5}{6} - 3\frac{1}{4}$
11. $2\frac{2}{5} - 1\frac{1}{4}$
12. $4\frac{7}{10} - 3\frac{1}{3}$

13. $7\frac{3}{5} - 4\frac{5}{6}$
14. $9\frac{3}{8} - 5\frac{1}{2}$
15. $8\frac{1}{9} - 5\frac{1}{3}$

16. $5\frac{1}{6} - 1\frac{2}{3}$
17. $5\frac{5}{6} - 3\frac{1}{3}$
18. $7\frac{2}{3} - 3\frac{5}{6}$

19. $8\frac{4}{7} - 4\frac{3}{4}$
20. $9\frac{3}{4} - 1\frac{1}{5}$
21. $5\frac{1}{2} - 2\frac{1}{3}$

FRACTION WORD PROBLEMS

Solve and reduce answers to lowest terms.

1. Sal works for a movie theater and sells candy by the pound. Her first customer bought $1\frac{1}{3}$ pounds of candy, the second bought $\frac{3}{4}$ of a pound, and the third bought $\frac{4}{5}$ of a pound. How many pounds did she sell to the first three customers?

2. Beth has a bread machine that makes a loaf of bread that weighs $1\frac{1}{2}$ pounds. If she makes a loaf of bread for each of her three sisters, how many pounds of bread will she make?

3. A farmer hauled in 120 bales of hay. Each of his cows ate $1\frac{1}{4}$ bales. How many cows did the farmer feed?

4. John was competing in a 1000 meter race. He had to pull out of the race after running $\frac{3}{4}$ of it. How many meters did he run?

5. Tad needs to measure where the free-throw line should be in front of his basketball goal. He knows his feet are $1\frac{1}{8}$ feet long and the free-throw line should be 15 feet from the backboard. How many toe-to-toe steps does Tad need to take to mark off 15 feet?

6. A chemical plant takes in $5\frac{1}{2}$ million gallons of water from a local river and discharges $3\frac{2}{3}$ million back into the river. How much water does not go back into the river?

7. In January, Jeff filled his car with $11\frac{1}{2}$ gallons of gas the first week, $13\frac{1}{3}$ gallons the second week, $12\frac{1}{4}$ gallons the third week, and $10\frac{1}{5}$ gallons the forth week of January. How many gallons of gas did he buy in January?

8. Martin makes sandwiches for his family. He has $8\frac{1}{4}$ ounces of sandwich meat. If he divides the meat equally to make $4\frac{1}{2}$ sandwiches, how much meat will each sandwich have?

9. The company water cooler started with $4\frac{1}{3}$ gallons of water. Employees drank $2\frac{3}{4}$ gallons. How many gallons were left in the cooler?

10. Rita bought $\frac{1}{4}$ pound hamburger patties for her family reunion picnic. She bought 50 patties. How many pounds of hamburgers did she buy?

CHAPTER REVIEW

Simplify.

1. $\dfrac{15}{6}$ _____

2. $\dfrac{24}{4}$ _____

3. $\dfrac{20}{15}$ _____

4. $\dfrac{14}{3}$ _____

Reduce.

5. $\dfrac{9}{27}$ _____

6. $\dfrac{4}{16}$ _____

7. $\dfrac{8}{12}$ _____

8. $\dfrac{12}{18}$ _____

Change to an improper fraction.

9. $5\dfrac{1}{10}$ _____

10. 7 _____

11. $3\dfrac{3}{5}$ _____

12. $6\dfrac{2}{3}$ _____

Add and simplify.

13. $\dfrac{5}{6} + \dfrac{7}{9}$ _____

14. $7\dfrac{1}{2} + 3\dfrac{3}{8}$ _____

15. $4\dfrac{4}{15} + \dfrac{1}{5}$ _____

16. $\dfrac{1}{7} + \dfrac{3}{7}$ _____

Subtract and simplify.

17. $10 - 5\dfrac{1}{8}$ _____

18. $3\dfrac{1}{3} - \dfrac{3}{4}$ _____

19. $9\dfrac{3}{4} - 2\dfrac{3}{8}$ _____

20. $6\dfrac{1}{5} - 1\dfrac{3}{10}$ _____

Multiply and simplify.

21. $1\dfrac{1}{3} \times 3\dfrac{1}{2}$ _____

22. $5\dfrac{3}{7} \times \dfrac{7}{8}$ _____

23. $4\dfrac{4}{6} \times 1\dfrac{5}{7}$ _____

24. $\dfrac{2}{3} \times \dfrac{5}{6}$ _____

Divide and simplify.

25. $\dfrac{1}{2} \div \dfrac{4}{5}$ _____

26. $6\dfrac{6}{7} \div 2\dfrac{2}{3}$ _____

27. $3\dfrac{5}{6} \div 11\dfrac{1}{2}$ _____

28. $1\dfrac{1}{5} \div 3\dfrac{1}{5}$ _____

29. Which fraction is between $\dfrac{1}{5}$ and $\dfrac{1}{2}$?

   A. $\dfrac{1}{3}$ B. $\dfrac{2}{3}$ C. $\dfrac{5}{6}$ D. $\dfrac{7}{15}$

30. Which fraction is between $\dfrac{1}{3}$ and $\dfrac{5}{9}$?

   A. $\dfrac{1}{5}$ B. $\dfrac{5}{6}$ C. $\dfrac{2}{9}$ D. $\dfrac{11}{18}$

31. Which fraction is between $\dfrac{1}{5}$ and $\dfrac{3}{7}$?

   A. $\dfrac{2}{5}$ B. $\dfrac{1}{7}$ C. $\dfrac{1}{2}$ D. $\dfrac{3}{4}$

Find the greatest common factor for the following sets of numbers.

32. 9 and 15 _____

33. 12 and 16 _____

34. 10 and 25 _____

35. 8 and 24 _____

Find the least common multiple for the following sets of numbers.

36. 8 and 12 _____

37. 5 and 9 _____

38. 4 and 10 _____

39. 6 and 8 _____

40. Mrs. Tate brought $5\frac{1}{2}$ pounds of candy to divide among her 22 students. If the candy was divided equally, how many pounds of candy did each student receive?

_____

41. Dorothy used $1\frac{1}{5}$ yards of material to recover one dining room chair. How much material would she need to recover all eight chairs?

_____

42. The square tiles in Mr. Cooke's math classroom measure $2\frac{1}{4}$ feet across. The students counted that the classroom was $5\frac{1}{3}$ tiles wide. How wide is Mr. Cooke's classroom?

_____

43. The Carlyle family is hiking a $23\frac{1}{3}$ mile trail. The first day, they hiked $16\frac{1}{2}$ miles. How much further do they have to go to complete the trail?

_____

44. Jena walked $\frac{1}{5}$ of a mile to a friend's house, $1\frac{1}{3}$ miles to the store, and $\frac{3}{4}$ of a mile back home. How far did Jena walk?

_____

45. Corey used $2\frac{4}{5}$ gallons of paint to mark one mile of this year's spring road race. How many gallons will he use to mark the entire $6\frac{1}{4}$ mile course?

_____

46. Mia used $7\frac{1}{2}$ pounds of beef to make $22\frac{1}{2}$ pounds of lasagna. How much beef was in each pound of lasagna?

_____

47. Cameron is selling licorice by the foot. So far, he has sold $2\frac{1}{3}$ feet to Sara, $1\frac{3}{8}$ feet to Dan, and $\frac{3}{4}$ foot to Bobby. How many total feet has Cameron sold?

_____

48. Greta purchased a roll of ribbon totaling $25\frac{1}{3}$ feet. She used $19\frac{1}{4}$ feet of the ribbon decorating for Jack's party. How many feet of ribbon remain on the roll?

_____

49. Yvonne biked for $1\frac{1}{4}$ hours and traveled $12\frac{1}{2}$ miles. How many miles per hour did she average?

_____

50. Ryan's 5 cats each eat $\frac{3}{4}$ cup of food every day. What is the total amount of food his cats eat each day?

_____

51. Hans juiced $\frac{1}{4}$ of a lemon to make 1 cup of lemonade. How many lemons should he juice to make $5\frac{1}{2}$ cups of lemonade?

_____

CHAPTER TEST

1. Ken had $90\frac{1}{2}$ feet of rope in his garage. He needed $73\frac{1}{3}$ feet to replace a rotted pulley rope. How many feet of rope did Ken have left?

   A. $17\frac{1}{5}$
   B. $17\frac{2}{5}$
   C. $18\frac{1}{6}$
   D. $17\frac{1}{6}$

2. Multiply: $2\frac{3}{4} \times 1\frac{1}{5}$

   A. $2\frac{3}{20}$
   B. $2\frac{7}{24}$
   C. $3\frac{3}{10}$
   D. $6\frac{3}{5}$

3. Which fraction is greater than $\frac{3}{5}$?

   A. $\frac{7}{8}$
   B. $\frac{1}{3}$
   C. $\frac{4}{15}$
   D. $\frac{3}{8}$

4. Holly bought 10 yards of fabric to recover 6 dining room chairs. Each chair took $1\frac{1}{4}$ yards. How much fabric did she have left?

   A. $2\frac{1}{2}$
   B. 4
   C. $4\frac{4}{5}$
   D. $7\frac{1}{2}$

5. $6\frac{1}{4}$
   $-3\frac{7}{8}$

   A. $2\frac{3}{8}$
   B. $2\frac{1}{2}$
   C. $2\frac{5}{8}$
   D. $3\frac{5}{8}$

6. Which fraction is between $\frac{1}{10}$ and $\frac{1}{1000}$?

   A. $\frac{2}{10}$
   B. $\frac{3}{1000}$
   C. $\frac{1}{100}$
   D. $\frac{101}{1000}$

7. $2\frac{5}{8}$
   $5\frac{2}{3}$
   $+\ \frac{1}{2}$

   A. $7\frac{1}{3}$
   B. $7\frac{8}{13}$
   C. $8\frac{1}{2}$
   D. $8\frac{19}{24}$

8. $1\frac{2}{4} \div 2\frac{5}{8}$

   A. $\frac{1}{2}$
   B. $\frac{2}{3}$
   C. $\frac{7}{8}$
   D. $1\frac{1}{2}$

9. What is the greatest common factor of 36 and 120?

   A. 2
   B. 6
   C. 12
   D. 18

10. What is the least common multiple of 6 and 8?

    A. 2
    B. 16
    C. 24
    D. 48

11.

    According to the sale ad above, how much could you save on a board game regularly priced at $12.00?

    A. $ 3.00
    B. $ 6.00
    C. $ 9.00
    D. $15.00

12. Which fraction is the greatest?

    A. $\frac{1}{6}$
    B. $\frac{2}{3}$
    C. $\frac{3}{5}$
    D. $\frac{9}{15}$

13. Which fraction is the smallest?

    A. $\frac{1}{2}$
    B. $\frac{2}{9}$
    C. $\frac{4}{7}$
    D. $\frac{1}{3}$

14. Henderson Middle School had 480 students in attendance last Monday, and $\frac{3}{5}$ of the students bought a hot lunch in the cafeteria. How many students bought a hot lunch?

    A. 96
    B. 160
    C. 288
    D. 800

15. Jerome bought $1\frac{1}{2}$ pounds of jelly beans, $\frac{5}{8}$ pound of gum balls, and $2\frac{3}{4}$ pounds of chocolate. How many total pounds of candy did he buy?

    A. $3\frac{7}{8}$ pounds
    B. $4\frac{3}{4}$ pounds
    C. $4\frac{7}{8}$ pounds
    D. 5 pounds

16. Kevin needs $3\frac{1}{3}$ yards of streamers to decorate each table. How many tables will 10 yards of streamers decorate?

    A. 3
    B. $3\frac{1}{3}$
    C. 10
    D. $66\frac{2}{3}$

17. Each lap around the lake is $\frac{4}{5}$ of a mile. How many miles did Teri run if she ran $4\frac{1}{2}$ laps?

    A. $3\frac{3}{5}$
    B. $3\frac{1}{5}$
    C. $4\frac{2}{5}$
    D. $5\frac{5}{8}$

Chapter 4

DECIMALS

ROUNDING DECIMALS

EXAMPLE:

Hundreds
Tens
Ones
Tenths
Hundredths
Thousandths
Ten Thousandths

5 6 8 . 4 5 8 7

Consider the number 568.4587 shown with the place values labeled to the left. To round to a given place value, first find the place value in the decimal. Then look to the digit on the right. If the digit on the right is 5 or greater, INCREASE BY ONE the place value you are rounding to. All the digits to the right of the given place value are dropped if the place value is after the decimal point. If the digit on the right is LESS THAN 5, leave the place value the same. All the digits to the right of the given place value are dropped if the place value you are rounding to is after the decimal point.

Round the number 568.4587 to the nearest:

Tenth	568.5
Hundredth	568.46
Thousandth	568.459

Note: The decimal point is never moved when rounding.

Round to the nearest tenth.

1. 45.58 _____
2. 17.041 _____
3. 5.284 _____
4. 0.618 _____
5. 4.543 _____
6. 91.385 _____
7. 25.483 _____
8. 9.022 _____
9. 0.349 _____
10. 0.854 _____
11. 21.028 _____
12. 1.957 _____

Round to the nearest hundredth.

13. 10.049 _____
14. 0.753 _____
15. 6.432 _____
16. 32.896 _____
17. 6.354 _____
18. 5.729 _____
19. 13.006 _____
20. 9.965 _____
21. 0.874 _____
22. 31.456 _____
23. 4.9571 _____
24. 8.6274 _____

Round to the nearest thousandth.

25. 9.8457 _____
26. 12.2854 _____
27. 0.8542 _____
28. 8.00295 _____
29. 7.1546 _____
30. 5.1238 _____
31. 4.0782 _____
32. 0.9984 _____
33. 5.2171 _____
34. 6.5832 _____
35. 2.7153 _____
36. 8.39548 _____

READING AND WRITING DECIMALS

EXAMPLE: Write the number 506.402 in words.

(The digits have been lined up with their place value names in the box on the right.)

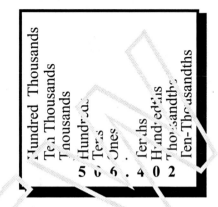

Step 1 **Write the number to the left of the decimal point:** Five hundred six

Step 2 **Write the word "and" where the decimal point is:** Five hundred six and

Step 3 **Write the numbers after the decimal point and add the name of the column of the last digit:** Five hundred six and four hundred two thousandths.

Note: Only use the word "and" where the decimal point comes. NEVER write three hundred and eight for the number 308.

Write the following numbers in words.

1. 951.32 _____
2. 542.236 _____
3. 10.058 _____
4. 901.47 _____
5. 5.0247 _____
6. 54.0587 _____
7. 23.503 _____
8. 65.0245 _____
9. 7.653 _____
10. 325.473 _____
11. 80.057 _____
12. 9.548 _____
13. 570.007 _____
14. 95.0345 _____

Use digits to write each of the following whole numbers and decimals. Remember the word "and" denotes a decimal point.

1. _____ twenty-two and four hundred ninety-five thousands

2. _____ six and seventy-four ten-thousandths

3. _____ one hundred sixty-five and thirty-four hundredths

4. _____ four thousand five hundred six and nineteen thousandths

5. _____ forty and three hundred ninety thousandths

6. _____ six hundred six and six hundred six thousandths

7. _____ nine and thirty-four hundredths

8. _____ fifty-five ten-thousandths

9. _____ fifteen thousand, three hundred sixty-seven and two tenths

10. _____ seven thousand eighty-nine and fifty-one thousandths

11. _____ one hundred eighteen thousand, two hundred five and ninety-eight hundredths

12. _____ seven and three hundred sixty-five ten-thousandths

13. _____ five hundred seventy and four hundredths

14. _____ seventeen and nine thousandths

15. _____ six and fourteen hundredths

16. _____ fifty thousand, four hundred twenty and nine hundred seventy-four thousandths

17. _____ eight hundred fifty-one and eighty-six ten-thousandths

18. _____ nineteen thousand, one hundred five and seven thousandths

19. _____ one and three thousand, four hundred seventy-five ten-thousandths

20. _____ fifty-two and five hundred nine thousandths

21. _____ twenty-two ten-thousandths

22. _____ six thousand, one hundred ninety-five ten-thousandths

ADDING DECIMALS

EXAMPLE: Find $0.9 + 2.5 + 63.17$

Step 1 When you add decimals, first arrange the numbers in columns with the decimal points under each other.

$$\begin{array}{r} 0.9 \\ 2.5 \\ +\ 63.17 \end{array}$$

Step 2 Add 0's here to keep your columns straight

Step 3 Start at the right and add each column. Remember to carry when necessary. Bring down the decimal point.

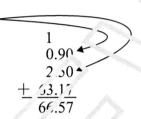

$$\begin{array}{r} 1\ \ \ \ \\ 0.90 \\ 2.50 \\ +\ 63.17 \\ \hline 66.57 \end{array}$$

Add. Be sure to write the decimal point in your answer.

1. $5.3 + 6.02 + 0.73$
2. $0.235 + 6.2 + 3.27$
3. $7.542 + 10.5 + 4.57$
4. $\$5.87 + \$7.52$
5. $\$4.68 + \$9.47$

6. $5.08 + 11.2 + 6.073$
7. $5.14 + 2.3 + 5.097$
8. $4.9 + 15.71 + 0.254$
9. $\$3.75 + \$18.90$
10. $\$64.95 + \$4.63$

11. $1.25 + 5.1 + 10.007$
12. $15.4 + 5.074 + 3.15$
13. $45.23 + 9.5 + 0.695$
14. $\$8.53 + \$12.50$
15. $\$6.87 + \$27.23$

16. $0.23 + 5.9 + 12$
17. $8.5784 + 10.05$
18. $55.7 + 205.952$
19. $\$98.45 + \$8.89$
20. $\$7.77 + \$11.19$

SUBTRACTING DECIMALS

EXAMPLE: Find 14.9 − 0.007

Step 1 When you subtract decimals, arrange the numbers in columns with the decimal points under each other.

$$\begin{array}{r} 14.9 \\ -\ 0.007 \end{array}$$

Step 2 You must fill in the empty places with 0's so that both numbers have the same number of digits after the decimal point.

$$\begin{array}{r} 14.900 \\ -\ 0.007 \end{array}$$

Step 3 Start at the right and subtract each column. Remember to borrow when necessary.

$$\begin{array}{r} 14.\overset{89}{9}\overset{1}{0}0 \\ -\ 0.007 \\ \hline 14.893 \end{array}$$

Subtract. Be sure to write the decimal point in your answer.

1. 5.25 − 4.7
2. 23.657 − 9.83
3. $56.54 − $17.02
4. $294.78 − $80.99
5. $70.00 − $68.99

6. 58.6 − 9.153
7. 405.97 − 7.325
8. $40.09 − $9.99
9. $115.45 − $4.79
10. $45.18 − $23.65

11. 12.26 − 7.32
12. 19.2 − 8.63
13. 8.125 − 5.096
14. $14.32 − $0.58
15. $30.00 − $22.95

16. 15.789 − 6.32
17. 478.65 − 99.2
18. $15.45 − $8.58
19. 102.5 − 1.079
20. 7.054 − 3.009

DETERMINING CHANGE

EXAMPLE: Jamie bought 2 t-shirts for $13.95 each and paid $1.68 sales tax. How much change should Jamie get from a $50.00 bill?

Step 1: Find the total cost of items and tax.

$13.95
13.95
+ 1.68
$29.58

Step 2: Subtract the total cost from the amount of money given.

$50.00
− 29.58
Change ⟶ $20.42

Find the correct change for each of the following problems.

1. Kenya bought a leather belt for $22.89 and a pair of earrings for $4.69. She paid $1.38 sales tax. What was her change from $30.00?

2. Mark spent a total of $78.42 on party supplies. What was his change from a $100.00 bill?

3. The Daniels spent $42.98 at a steak restaurant. How much change did they receive from $50.00?

4. Myra bought a sweater for $49.95 and a dress for $85.89. She paid $9.51 in sales tax. What was her change from $150.00?

5. Roland bought a calculator for $22.78 and an extra battery for $5.69. He paid $1.56 sales tax. What was his change from $40.00?

6. For lunch, Daul purchased 2 hotdogs for $1.09 each, a bag of chips for $0.89, and a large drink for $1.59. He paid $0.18 sales tax. What was his change from $10.00?

7. Geri bought a dining room set for $2,265.99. She paid $135.96 sales tax. What was her change from $2,500.00?

8. Juan purchased a bag of dog food for $5.89, a leash for $11.88, and a dog collar for $4.75. The sales tax on the purchase was $1.13. How much change did he get back from 25.00?

9. Maxine bought a blouse for $15.46 and a shirt for $23.58. She paid $2.12 sales tax. What was her change from $50.00?

10. Jackie paid for four houseplants that cost $4.95 each. She paid $1.19 sales tax. How much change did she receive from 30.00?

11. Bo spent a total of $13.59 on school supplies. How much change did he receive from $14.00?

12. Fran bought 4 packs of candy on sale for 2 for $0.99 and 2 sodas for $0.65 each. She paid $0.13 sales tax. What was her change from a $5.00 bill?

MULTIPLICATION OF DECIMALS

EXAMPLE: 56.2 × 0.17

Step 1 Set up the problem as if you were multiplying whole numbers.

```
  56.2
× 0.17
```

Step 2 Multiply as if you were multiplying whole numbers.

```
   4 1
   56.2    ← 1 number after the decimal point
 × 0.17    ← + 2 numbers after the decimal point
   3934      3 numbers after the decimal point
   562
   9.554
```

Step 3 Count how many numbers are after the decimal points in the problem. In this problem 2, 1 and 7 come after the decimal points, so the answer must also have three numbers after the decimal point.

Multiply.

1. 15.2 × 3.5
2. 9.54 × 5.3
3. 5.72 × 6.3
4. 4.8 × 3.2

5. 45.8 × 2.2
6. 4.5 × 7.1
7. 0.052 × 0.33
8. 4.12 × 6.8

9. 23.65 × 9.2
10. 1.54 × 0.43
11. 0.47 × 6.1
12. 1.3 × 1.57

13. 10.4 × 0.5
14. 0.87 × 3.21
15. 5.94 × 0.65
16. 7.8 × 0.23

MORE MULTIPLYING DECIMALS

EXAMPLE: Find 0.007 × 0.125

Step 1 Multiply as you would whole numbers.

$$\begin{array}{r} 0.007 \\ \times\ 0.125 \\ \hline 0.000875 \end{array}$$ ← 3 numbers after the decimal point
← +3 numbers after the decimal point
← 6 numbers after the decimal point

Step 2 Count how many numbers are behind decimal points in the problem. In this case, 6 numbers come after decimal points in the problem, so there must be 6 numbers after the decimal point in the answer. In this problem, 0's needed to be written in the answer **in front of** the 8 so there can be 6 numbers after the decimal point.

Multiply. Write in zeros as needed. Round dollar figures to the nearest penny.

1. 0.123 × .45
2. 0.004 × 10.31
3. 1.54 × 1.1
4. 10.03 × 0.45

5. 9.45 × 0.8
6. 50.4 × 0.06
7. 5.003 × 0.009
8. $9.99 × 0.06

9. 6.09 × 5.3
10. $22.00 × 0.075
11. 5.914 × 0.02
12. 4.95 × 0.23

13. 6.98 × 0.02
14. 3.12 × 0.08
15. 7.158 × 0.09
16. 0.0158 × 0.32

GROSS PAY

Gross pay is the amount you earn before taxes, insurance, and other deductions are taken out.

EXAMPLE: Codie earns $8.50 per hour. Last week he worked 38 hours. What was his gross pay?

Solution: Multiply the pay per hour by the number of hours worked.

$$\begin{array}{r} \$8.50 \\ \times\ 38 \\ \hline 6800 \\ 2550 \\ \hline \$323.00 \end{array}$$

Find the gross pay (total earnings before deductions) in each of the following problems.

1. Ron earns $8.25 per hour and works 35 hours each week. How much does he earn per week?

2. Casie earns $13.00 per hour and worked 40 hours last week. How much did she earn last week?

3. Maria earns $6.75 an hour at her part-time job. Last week she worked 15 hours. How much did she earn?

4. Roby worked 22.5 hours last week and he earns $8.60 per hour. How much was his gross pay?

5. Tikki worked 11 hours at her job that pays $9.15 per hour. How much did she earn?

6. Murray worked 35 hours last week and makes $6.45 per hour. What was his gross pay?

7. Paula's job pays $12.00 per hour. Last week she worked 11.25 hours. How much did she earn last week?

8. Taylor earns $6.50 per hour working in a fast food restaurant. If he works 23 hours per week, what is his gross pay per week?

9. Mark earns $9.50 per hour painting houses. Last week he worked 36 hours. How much did he earn?

10. Kirby works in a greenhouse for $8.75 per hour. He works 40 hours per week. What is his gross pay each week?

11. Julie earns $25.00 per hour teaching tennis part time. If she works 8.5 hours in a week, how much will she earn?

12. Yvonne works in a boutique for 25 hours per week. The boutique pays her $7.15 per hour. How much does she earn each week?

13. Rick works for a landscape architect for $9.45 per hour. If he puts in 32 hours per week, how much will he earn?

DIVISION OF DECIMALS BY WHOLE NUMBERS

EXAMPLE: $52.26 \div 6$

Step 1 Copy the problem as you would for whole numbers. Copy the decimal point directly above in the place for the answer.

$$6\overline{)52.26}$$

Step 2 Divide the same way as you would with whole numbers.

$$\begin{array}{r} 8.71 \\ 6\overline{)52.26} \\ -48 \\ \hline 4\,2 \\ -4\,2 \\ \hline 0\,6 \\ -6 \\ \hline 0 \end{array}$$

Divide. Remember to copy the decimal point directly above the place for the answer.

1. $42.75 \div 3$
2. $74.16 \div 6$
3. $81.50 \div 25$
4. $82.46 \div 14$

5. $12.50 \div 2$
6. $224.64 \div 52$
7. $183.04 \div 52$
8. $281.52 \div 23$

9. $72.36 \div 4$
10. $379.5 \div 15$
11. $52.25 \div 21$
12. $40.375 \div 19$

13. $102.5 \div 5$
14. $113.4 \div 9$
15. $585.14 \div 34$
16. $93.6 \div 24$

CHANGING FRACTIONS TO DECIMALS

EXAMPLE: Change $\frac{1}{8}$ to a decimal.

Step 1 To change a fraction to a decimal, simply divide the top number by the bottom number.

$$8\overline{)1}$$

Step 2 Add a decimal point and a 0 after the 1 and divide.

$$\begin{array}{r} 0.1 \\ 8\overline{)1.0} \\ -8 \\ \hline 2 \end{array}$$

Step 3 Continue adding 0's and dividing until there is no remainder.

$$\begin{array}{r} 0.125 \\ 8\overline{)1.000} \\ -8 \\ \hline 20 \\ -16 \\ \hline 40 \\ -40 \\ \hline 0 \end{array}$$

In some problems, the number after the decimal point begins to repeat. Take, for example, the fraction $\frac{4}{11}$. $4 \div 11 = 0.363636$ and the 36 keeps repeating forever. To show that the 36 repeats, simply write a bar above the numbers that repeat, $0.\overline{36}$.

Change the following fractions to decimals.

1. $\frac{4}{5}$
2. $\frac{2}{3}$
3. $\frac{1}{2}$
4. $\frac{5}{9}$

5. $\frac{1}{10}$
6. $\frac{5}{8}$
7. $\frac{5}{6}$
8. $\frac{1}{6}$

9. $\frac{3}{5}$
10. $\frac{7}{10}$
11. $\frac{4}{11}$
12. $\frac{1}{9}$

13. $\frac{7}{9}$
14. $\frac{9}{10}$
15. $\frac{1}{4}$
16. $\frac{3}{8}$

17. $\frac{3}{16}$
18. $\frac{3}{4}$
19. $\frac{8}{9}$
20. $\frac{5}{12}$

CHANGING MIXED NUMBERS TO DECIMALS

If there is a whole number with a fraction, write the whole number to the left of the decimal point. Then change the fraction to a decimal.

EXAMPLES: $4\frac{1}{10} = 4.1$ $16\frac{2}{3} = 16.\overline{6}$ $12\frac{7}{8} = 12.875$

Change the following mixed numbers to decimals.

1. $5\frac{2}{3}$
2. $8\frac{5}{11}$
3. $15\frac{3}{5}$
4. $13\frac{2}{3}$
5. $30\frac{1}{3}$

6. $3\frac{1}{2}$
7. $1\frac{7}{8}$
8. $4\frac{9}{100}$
9. $6\frac{4}{5}$
10. $13\frac{1}{2}$

11. $12\frac{4}{5}$
12. $11\frac{5}{8}$
13. $7\frac{1}{4}$
14. $12\frac{1}{3}$
15. $1\frac{5}{9}$

16. $2\frac{3}{4}$
17. $10\frac{1}{10}$
18. $20\frac{2}{5}$
19. $4\frac{9}{10}$
20. $5\frac{4}{11}$

CHANGING DECIMALS TO FRACTIONS

EXAMPLE: 0.25

Step 1: Copy the decimal without the point. This will be the top number of the fraction.

$\frac{25}{\Box}$

Step 2: The bottom number is a 1 with as many 0's after it as there are digits in the top number.

$\frac{25}{100}$ ← Two digits / Two 0's

$\frac{25}{100} = \frac{1}{4}$

Step 3: You then need to reduce the fraction.

EXAMPLES: $.2 = \frac{2}{10} = \frac{1}{5}$ $.65 = \frac{65}{100} = \frac{13}{20}$ $.125 = \frac{125}{1000} = \frac{1}{8}$

Change the following decimals to fractions.

1. .55
2. .6
3. .12
4. .9
5. .75
6. .82
7. .3
8. .42
9. .71
10. .42
11. .56
12. .24
13. .35
14. .96
15. .125
16. .275

CHANGING DECIMALS WITH WHOLE NUMBERS TO MIXED NUMBERS

EXAMPLE: Change 14.28 to a mixed number.

Step 1: Copy the portion of the number that is whole.   14

Step 2: Change .28 to a fraction $14\frac{28}{100}$

Step 3: Reduce the fraction $14\frac{28}{100} = 14\frac{7}{25}$

Change the following decimals to mixed numbers.

1. 7.125
2. 99.5
3. 2.13
4. 5.1
5. 16.95
6. 3.625
7. 4.42
8. 15.84
9. 5.7
10. 45.425
11. 15.8
12. 8.16
13. 13.9
14. 32.65
15. 17.25
16. 9.82

DIVISION OF DECIMALS BY DECIMALS

EXAMPLE: 374.5 ÷ 0.07

Step 1 Copy the problem as you would for whole numbers.

$$0.07 \overline{) 374.5}$$

Step 2 You cannot divide by a decimal number. You must move the decimal point in the divisor 2 places to the right to make it a whole number. The decimal point in the dividend must also move to the right the same number of places. Notice you must add a 0 to the dividend.

Step 3 The problem now becomes 37450 ÷ 7. Copy the decimal point from the dividend straight above in the place for the answer.

Divide. Remember to move the decimal point.

1. 0.676 ÷ 0.013
2. 70.32 ÷ 0.08
3. $54.60 ÷ 0.84
4. $10.35 ÷ 0.45

5. 18.46 ÷ 1.2
6. 14.6 ÷ 0.002
7. $125.25 ÷ 0.75
8. $33.00 ÷ 1.65

9. 154.08 ÷ 1.8
10. 0.4374 ÷ 0.003
11. 292.2 ÷ 0.25
12. 6.375 ÷ 0.3

13. 4.8 ÷ 0.08
14. 1.2 ÷ 0.024
15. 15.725 ÷ 3.7
16. $167.50 ÷ 0.25

BEST BUY

When products come in different sizes, you need to figure out the cost per unit to see which is the best buy. Often the box marked "economy size" is not really the best buy.

EXAMPLE:
Smithfield's Instant Coffee comes in three sizes. Which one has the best cost per unit? The coffee comes in 8, 12, and 16 ounce sizes. To figure the least cost per unit, you need to see how much each unit, in this case ounce, costs in each size. If 8 oz of coffee costs $3.60, then 1 oz costs $3.60 ÷ 8 or $.45. $.45 is the unit cost, the cost of 1 oz. We need to figure the unit cost for each size:

$3.60 ÷ 8 = $.45
$5.52 ÷ 12 = $.46
$7.44 ÷ 16 = $.465

The 8 oz size is the best one to buy because it has the lowest cost per unit.

Figure the unit cost of each item in each question below to find the best buy. Underline the answer.

1. Which costs the most per ounce, 60 oz of peanut butter for $5.40, 28 oz for $2.24, or 16 oz for $1.76?

2. Which is the least per pound, 5 lb of chicken for $9.45, 3 lb for $5.97, or 1 lb for $2.05?

3. Which costs the most per disk, a 10 pack of 3½ inch floppy disks for $5.99, a 25 pack for $12.50, or a 50 pack for $18.75?

4. Which is the best buy, 6 ballpoint pens for $4.80 or 8 for $6.48?

5. Which costs the least per ounce, a 20 oz soda for $0.60, 68 oz for $2.38, or 100 oz for $3.32?

6. Which costs more, oranges selling at 3 for $1.00 or oranges selling 4 for $1.36?

7. Which is the best buy, 1 roll of paper towels for $2.13, 3 rolls for $5.88, or 15 rolls for $29.55?

8. Which costs the most per tablet, 50 individually wrapped pain reliever tablets for $9.50, 100 tablets in a bottle for $6.52, or 500 tablets in a bottle for $12.73?

9. Which costs the least per can, a 24 pack of cola for $5.52, a 12 pack of cola for $2.64, or a 6 pack of cola for $1.35?

10. Which costs less per bag, 18 tea bags for $2.70 or 64 tea bags for $9.28?

11. Which is the best buy, a 3 pack of correction fluid for $2.97 or a 12 pack for $11.76?

12. Which is the least per roll, a roll of masking tape for $2.45, a 3-roll pack for $7.38, or a 12-roll pack for $29.16?

BEST BUY SAVINGS

EXAMPLE:

Item:	Store A	Store B
gallon of paint	$15.99	$17.89

To find the savings, subtract the lower price from the higher price.

$$\begin{array}{r} \$17.89 \\ -\ \$15.99 \\ \hline \$\ 1.90 \end{array}$$

savings ⟶ $1.90

Find the amount of savings by choosing the lower price in each of the following. Assume items are identical in each store.

Item	Store A	Store B	Savings
1. 6 pack of soda	$2.49	$3.12	_____
2. alarm clock	$15.95	$12.79	_____
3. space heater	$23.85	$25.97	_____
4. sports car	$25,890	$29,489	_____
5. video game	$19.99	$18.77	_____
6. garden shovel	$7.89	$15.39	_____
7. mountain bike	$596	$603	_____
8. goldfish	$0.89	$0.56	_____
9. large screen TV	$1579	$1245	_____
10. roller blades	$116.95	$121.79	_____
11. cotton shorts	$16.99	$12.79	_____
12. oriental rug	$152.86	$169.45	_____
13. wrist watch	$19.59	$16.38	_____
14. gold hoop earrings	$48.75	$56.79	_____

ORDERING DECIMALS

EXAMPLE: Order the following decimals from greatest to least.

.3, .029, .208, .34

Step 1 Arrange numbers with decimal points directly under each other.

.3
.029
.208
.34

Step 2 Fill in with 0's so they all have the same number of places after the decimal point.

Read the numbers as if the decimal point wasn't there.

.300
.029 ← Least
.208
.340 ← Greatest

Answer: .34, .3, .208, .029

Order each set of decimals below from greatest to least.

1. .075, .705, .7, .75
2. .5, .56, .65, .06
3. .9, .09, .099, .95
4. .6, .59, .06, .66
5. .3, .303, .03, .33
6. .02, .25, .205, .5
7. .004, .44, .045, .4

8. .52, .905, .509, .099
9. .1, .01, .11, .111
10. .87, .078, .78, .8
11. .41, .45, .409, .49
12. .754, .7, .74, .75
13. .63, .069, .07, .06
14. .25, .275, .208, .027

Order each set of decimals below from least to greatest.

15. .055, .5, .59, .05
16. .7, .732, .74, .72
17. .04, .48, .048, .408
18. .9, .905, .95, .09
19. .19, .09, .9, .1
20. .21, .02, .021, .2
21. .038, .3, .04, .38

22. .695, .59, .065, .69
23. .08, .88, .808, .008
24. .015, .05, .105, .15
25. .4, .407, .47, .047
26. .632, .63, .603, .62
27. .02, .022, .222, .20
28. .541, .54, .504, .5

DECIMAL WORD PROBLEMS

1. Micah can have his oil changed in his car for $19.99, or he can buy the oil and filter and change it himself for $8.79. How much would he save by changing the oil himself?

2. Megan bought 5 boxes of cookies for $3.75 each. How much did she spend?

3. Will subscribes to a monthly auto magazine. His one year subscription cost $29.97. If he pays for the subscription in 3 equal installments, how much is each payment?

4. Pat purchases 2.5 pounds of hamburger at $0.98 per pound. What is the total cost of the hamburger?

5. The White family took $650 cash with them on vacation. At the end of their vacation, they had $4.67 left. How much cash did they spend on vacation?

6. Acer Middle School spent $1443.20 on 55 math books. How much did each book cost?

7. The Junior Beta Club needs to raise $1513.75 to go to a national convention. If they decide to sell candy bars at $1.25 each, how many will they need to sell to meet their goal?

8. Fleta owns a candy store. On Monday, she sold 6.5 pounds of chocolate, 8.34 pounds of jelly beans, 4.9 pounds of sour snaps, and 2.64 pounds of yogurt-covered raisins. How many pounds of candy did she sell total?

9. Randal purchased a rare coin collection for $1093.95. He sold it at auction for $2700. How much money did he make on the coins?

10. A leather jacket that normally sells for $259.99 is on sale now for $197.88. How much can you save if you buy it now?

11. At the movies, Gigi buys 0.6 pounds of candy priced at $2.10 per pound. How much did she spend on candy?

12. George has $6.00 to buy candy. If each candy bar costs $.60, how many bars can he buy?

13. While training for a marathon, Todd runs 12.5 miles on Monday, 15.6 miles on Tuesday, 19.25 miles on Wednesday, 20.45 miles on Thursday, and 13 miles on Friday. How many miles did he run Monday through Friday?

    _____

14. Thomas worked 35 hours and earned $288.75 in gross pay. How much does he make per hour?

    _____

15. Kathy earns a salary plus commission. Her base salary is $145.25 per week. If last week her gross pay was $384.12, how much of her pay was from commission?

    _____

16. A copy shop leases a copy machine for $256.25 per month plus $0.008 per copy made on the machine. If the copy shop makes 12,480 copies in a month, what will be the lease payment?

    _____

17. Mike purchased a CD for $12.95, batteries for $7.54, and earphones for $24.68. He paid $2.94 in sales tax. How much did he spend?

    _____

18. If gasoline costs $1.23 per gallon, how much would it cost to buy 12.5 gallons?

    _____

19. Kevin is saving money to buy a TV that sells for $350.75. He earns money by cutting his neighbors lawns for $15.25 per lawn. How many lawns must he cut before he will have enough money to buy the TV?

    _____

20. T-shirts were donated at no cost to the Barker Valley High School Band. The band then sold 278 of the t-shirts at $8.75 each to raise money. How much money did they raise?

    _____

21. Neil drives 23.48 miles to work each day and the same distance back home. How many total miles does he travel to work and back Monday through Friday?

    _____

22. When Jude travels on company business using his own car, his company reimburses him $0.26 per mile. How much will Jude get reimbursed if he travels 352.5 miles?

    _____

CHAPTER REVIEW

Round the following numbers to the nearest tenth.
1. 32.235 _____
2. 4.561 _____
3. 0.075 _____

Round the following numbers to the nearest hundredth.
4. 501.479 _____
5. 72.492 _____
6. 30.8472 _____

Round the following numbers to the nearest thousandth.
7. 2.0392 _____
8. 0.0057212 _____
9. 14.0851 _____

Write out in words.
10. 405.95 _____
_____

11. 5.047 _____
_____

12. 0.746 _____
_____
_____

Express in digits.
13. One hundred ninety-two and thirty-five hundredths _____

14. Two and twenty-nine hundredths _____

15. Seven thousand, four hundred two and three hundred sixty-three ten-thousandths _____

Add.
16. $12.589 + 5.62 + 0.9$ _____

17. $7.8 + 10.24 + 1.903$ _____

18. $152.64 + 12.3 + 0.024$ _____

Subtract.
19. $18.547 - 9.62$ _____

20. $1.85 - 0.693$ _____

21. $65.2 - 57.9$ _____

Multiply.
22. 4.58×0.025 _____

23. 0.879×1.7 _____

24. 30.7×0.004 _____

Divide.
25. $17.28 \div .054$ _____

26. $174.66 \div 1.23$ _____

27. $2.115 \div 9$ _____

Change to a fraction.
28. 0.55 _____
29. 0.54 _____
30. 0.32 _____

Change to a mixed number.

31. 7.375 _____

32. 9.6 _____

33. 13.25 _____

Change to a decimal.

34. $5\frac{3}{25}$ _____

35. $\frac{7}{100}$ _____

36. $10\frac{2}{3}$ _____

Put the following sets of decimals in order from GREATEST to LEAST.

37. 0.5, 0.55, 0.505, 0.05

_____

38. 0.24, 0.201, 0.022, 0.2

_____

39. 1.89, 1.08, 0.98, 0.9

_____

Put the following sets of decimals in order from LEAST to GREATEST.

40. 0.59, 0.5, 0.059, 0.509

_____

41. 0.19, 0.2, 0.109, 0.22

_____

42. 1.75, 0.79, 1.709, 1.8

_____

43. Super-X sells tires for $24.56 each. Save-Rite sells the identical tire for $21.97. How much can you save by purchasing a tire from Save-Rite?

_____

44. White rice sells for $5.64 for a 20 pound bag. A three pound bag costs $1.59. Which is the better buy?

_____

45. Daisy's Discount Mart sells 500 sheet packs of notebook paper for $4.00, 125 sheet packs for $1.17, and 200 sheet packs for $1.40. Which is the best buy?

_____

46. Xandra bought a mechanical pencil for $2.38, 3 pens for $0.89 each, and a pack of graph paper for $3.42. She paid $0.42 tax. What was her change from a ten dollar bill?

_____

47. Charlie makes $13.45 per hour repairing lawn mowers part time. If he worked 15 hours, how much was his gross pay?

_____

48. Gene works for his father sanding wooden rocking chairs. He earns $6.35 per chair. How many chairs does he need to sand in order to buy a portable radio/CD player for $146.05?

_____

49. Margo's Mint Shop has a machine that produces 4.35 pounds of mints per hour. How many pounds of mints are produced in each 8 hour shift?

_____

50. Carter's Junior High track team ran the first leg of a 400 meter relay race in 10.23 seconds, the second leg in 11.4 seconds, the third leg in 10.77 seconds, and the last leg in 9.9 seconds. How long did it take for them to complete the race?

_____

CHAPTER TEST

1. 0.45 written as a fraction is

    A. $\frac{9}{20}$
    B. $\frac{5}{9}$
    C. $\frac{4}{5}$
    D. $\frac{1}{45}$

2. $\frac{3}{8}$ written as a decimal is

    A. 0.375
    B. 0.38
    C. 0.0375
    D. 0.3

3. $4\frac{5}{6}$ written as a decimal is

    A. 0.456
    B. 4.56
    C. $4.8\overline{3}$
    D. $0.48\overline{3}$

4. 14.2 is the same as

    A. $\frac{142}{100}$
    B. $14\frac{1}{50}$
    C. $14\frac{1}{5}$
    D. $14\frac{1}{10}$

5. $40.2 + 4.02 + 200.4 + 40.02 =$

    A. 68.08
    B. 68.1
    C. 284.64
    D. 284.82

6. 508.2
   − 79.4

    A. 428.8
    B. 432.8
    C. 538.8
    D. 571.2

7. 0.05
   × 17.8

    A. 0.089
    B. 0.89
    C. 17.85
    D. 89

8. $1.008 \div 0.02$

    A. 5.04
    B. 50.4
    C. 504
    D. 5040

9. Monica's monthly salary is $2,100. Her deductions are $157.50 for FICA, $302 for federal income tax, and $57.80 for state income tax. What is her take-home pay?

    A. $1,682.70
    B. $1,617.30
    C. $1,582.69
    D. $1,582.70

10. Richard buys a camcorder for $229.95. Two months later, he sees the same camcorder on sale for $207.99. How much could he have saved if he waited to buy the camcorder on sale?

    A. $21.04
    B. $21.96
    C. $22.06
    D. $22.96

11. Tamara bought 2 movie tickets for $6.50 each, 2 colas for $1.25 each, and two bags of popcorn for $1.50 each. How much did Tamara spend in all?

    A. $9.25
    B. $15.50
    C. $18.50
    D. $18.75

12. Rosa worked 28 hours this week and was paid $5.60 per hour. What were her total earnings for the week?

    A. $ 15.68
    B. $ 22.40
    C. $ 33.60
    D. $156.80

13. Carl needs $231.75 to buy a new bike. If he earns $5.15 per hour net at a part-time job, how many hours would he have to work to earn enough to buy the bike?

    A. 43 hours
    B. 44.25 hours
    C. 45 hours
    D. 46.2 hours

14. Round the number 17.5694 to the nearest hundredth.

    A. 17
    B. 17.5
    C. 17.56
    D. 17.57

15. If 15 pencils cost $1.20, what is the cost of one pencil?

    A. 8¢
    B. 9¢
    C. 15¢
    D. 18¢

16. Which of the following is the correct way to spell out 801.35?

    A. eight hundred one and thirty-five hundredths
    B. eight hundred and one and thirty-five hundredths
    C. eight hundred one thirty-five
    D. eight hundred one and thirty-five

17. "Forty-five and twenty-nine thousands" expressed in digits is

    A. 0.4529
    B. 0.45029
    C. 45.29
    D. 45.029

18. Round 2.365 to the nearest tenth.

    A. 2.3
    B. 2.36
    C. 2.37
    D. 2.4

19. Round 45.871 to the nearest hundredth.

    A. 46.0
    B. 45.9
    C. 45.87
    D. 45.8

20. Which is the best buy?

    A. 3 candy bars for $0.96
    B. 4 candy bars for $1.00
    C. 5 candy bars for $1.20
    D. 6 candy bars for $2.10

21. Which set of decimals is in order from GREATEST to LEAST?

    A. .3, .36, .036, .06
    B. .95, .9, .095, .09
    C. .048, .4, .48, .5
    D. .2, .21, .02, .021

22. Which set of decimals is in order from LEAST to GREATEST?

    A. .64, .604, .064, .06
    B. .603, .75, .08, .098
    C. .5, .064, .751, .901
    D. .007, .069, .69, .7

23. Hanna bought 3 pairs of socks priced at 3 for $5.00 and shoes for $45.95. She paid $2.55 sales tax. How much change did she receive from $100.00?

    A. $36.50
    B. $46.50
    C. $51.50
    D. $53.50

Chapter 5: RATIOS, PROBABILITY, PROPORTIONS, AND SCALE DRAWINGS

RATIO PROBLEMS

In some word problems, you may be asked to express answers as a **ratio**. Ratios can look like fractions. Numbers must be written in the order they are requested. In the following problem, 8 cups of sugar is mentioned before 6 cups of strawberries. But in the question part of the problem, you are asked for the ratio of STRAWBERRIES to SUGAR. The amount of strawberries IS THE FIRST WORD MENTIONED so it must be the **top** number of the fraction. The amount of sugar, THE SECOND WORD MENTIONED, must be the **bottom** number of the fraction.

EXAMPLE: The recipe for jam requires 8 cups of sugar for every 6 cups of strawberries. What is the ratio of strawberries to sugar in the recipe?

First number requested $\quad\quad\dfrac{6 \text{ cups strawberries}}{8 \text{ cups sugar}}$
Second number requested

Answers may be reduced to lowest terms. $\dfrac{6}{8} = \dfrac{3}{4}$

Practice writing ratios for the following word problems and reduce to lowest terms. **DO NOT CHANGE ANSWERS TO MIXED NUMBERS. Ratios should be left in fraction form.**

1. Out of the 248 seniors, 112 are boys. What is the ratio of boys to the total number of seniors?

2. It takes 7 cups of flour to make 2 loaves of bread. What is the ratio of cups of flour to loaves of bread?

3. A skyscraper that stands 620 feet tall casts a shadow that is 125 feet long. What is the ratio of the shadow to the height of the skyscraper?

4. Jordan spent $45 on groceries. Of that total, $23 was for steaks. What is the ratio of steak cost to the total grocery cost?

5. The newborn weighs 6 pounds and is 22 inches long. What is the ratio of weight to length?

6. Jack paid $6.00 for 10 pounds of apples. What is the ratio of the price of apples to the pounds of apples?

7. Twenty boxes of paper weigh 520 pounds. What is the ratio of boxes to pounds?

8. Madison's flower garden measures 8 feet long by 6 feet wide. What is the ratio of length to width?

PROBABILITY

Probability is the chance something will happen. Probability is most often expressed as a fraction.

EXAMPLE 1: Billy had 3 red marbles, 5 white marbles, and 4 blue marbles on the floor. His cat came along and batted one marble under the chair. What is the **probability** it was a red marble?

Step 1: The number of red marbles will be the top number of the fraction. ⟶ $\frac{3}{12}$

Step 2 The total number of marbles is the bottom number of the fraction. ⟶ $\frac{3}{12}$

The answer may be expressed in lowest terms. $\frac{3}{12} = \frac{1}{4}$

EXAMPLE 2: Determine the probability that the pointer will stop on a shaded wedge or the number 1.

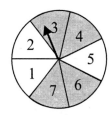

Step 1: Count the number of possible wedges that the spinner can stop on to satisfy the above problem. There are 5 wedges that satisfy it (4 shaded wedges and one number 1). The top number of the fraction is 5.

Step 2: Count the total number of wedges, 7. The bottom number of the fraction is 7.

The answer is $\frac{5}{7}$.

EXAMPLE 3: Refer to the spinner above. If the pointer stops on the number 7, what is the probability that it will **not** stop on 7 on the next spin?

Ignore the information that the pointer stopped on the number 7 on the previous spin. The probability of the next spin does not depend on the outcome of the previous spin. Simply find the probability that the spinner will **not** stop on 7.

In other words the question is asking: What is the probability the spinner will land on one of the other 6 wedges?

Therefore, the probability is $\frac{6}{7}$.

Find the probability of the following problems.

1. A computer chose a random number between 1 and 50. What is the probability of a person guessing the same number that the computer chose?

   _____

2. There are 24 candy-coated chocolate pieces in a bag. Eight have defects in the coating that can be seen only with close inspection. What is the probability of pulling out a defective piece without looking?

   _____

3. Seven sisters have to choose which day each will wash the dishes. They put equal size pieces of paper each labeled with a day of the week in a hat. What is the probability that the first sister who draws will choose a weekend day?

   _____

4. For his garden, Clay has a mixture of 12 white corn seeds, 24 yellow corn seeds, and 16 bi-color corn seeds. If he reaches for a seed without looking, what is the probability that Clay will plant a bi-color corn seed first?

   _____

5. Mom just got a new department store credit card in the mail. What is the probability that the last digit is an odd number?

   _____

6. Alex has a paper bag of cookies that includes 8 chocolate chip, 4 peanut butter, 6 butterscotch chip, and 12 ginger. Without looking, his friend John reaches in the bag for a cookie. What is the probability that the cookie is peanut butter?

   _____

7. An umpire at a little league baseball game has 14 balls in his pockets. Five of the balls are brand A, 6 are brand B, and 3 are brand C. What is the probability that the next ball he throws to the pitcher is a brand C ball?

   _____

8. What is the probability that the spinner arrow will land on an even number?

   _____

9. The spinner in the problem above stopped on a shaded wedge on the first spin and stopped on the number 2 on the second spin. What is the probability that it will not stop on a shaded wedge or on the 2 on the third spin?

   _____

10. A company is offering 1 grand prize, 3 second place prizes, and 25 third place prizes based on a random drawing of contest entries. If you entered one of the total 500 entries, what is the probability you will win a third place prize?

    _____

11. In the contest problem above, what is the probability that you will win the grand prize or a second place prize?

    _____

12. A box of a dozen donuts has 3 lemon cream-filled, 5 chocolate cream-filled, and 4 vanilla cream-filled. If the donuts look identical, what is the probability of picking a lemon cream-filled?

    _____

SOLVING PROPORTIONS

Two **ratios (fractions)** that are **equal** to each other are called **proportions**. For example, $\frac{1}{4} = \frac{2}{8}$. Read the following example to see how to find a number missing from a proportion.

EXAMPLE: $\frac{5}{15} = \frac{8}{x}$

Step 1 To find x, you first multiply the two numbers that are diagonal to each other. $15 \times 8 = 120$

Step 2 Then divide the product (120) by the other number in the proportion (5). $120 \div 5 = 24$

Therefore, $\frac{5}{15} = \frac{8}{24}$ $x = 24$

Practice finding the number missing from the following proportions. First, multiply the two numbers that are diagonal from each other. Then divide by the other number.

1. $\frac{2}{5} = \frac{6}{x}$

2. $\frac{9}{3} = \frac{x}{5}$

3. $\frac{x}{12} = \frac{3}{4}$

4. $\frac{7}{x} = \frac{3}{9}$

5. $\frac{12}{x} = \frac{2}{3}$

6. $\frac{12}{x} = \frac{4}{3}$

7. $\frac{27}{3} = \frac{x}{2}$

8. $\frac{1}{x} = \frac{3}{12}$

9. $\frac{15}{2} = \frac{x}{4}$

10. $\frac{7}{14} = \frac{x}{6}$

11. $\frac{5}{6} = \frac{10}{x}$

12. $\frac{4}{x} = \frac{3}{6}$

13. $\frac{x}{5} = \frac{9}{15}$

14. $\frac{9}{18} = \frac{x}{2}$

15. $\frac{5}{7} = \frac{35}{x}$

16. $\frac{x}{2} = \frac{8}{4}$

17. $\frac{15}{20} = \frac{x}{8}$

18. $\frac{x}{20} = \frac{5}{100}$

19. $\frac{4}{7} = \frac{x}{28}$

20. $\frac{7}{6} = \frac{42}{x}$

21. $\frac{x}{8} = \frac{1}{4}$

RATIO AND PROPORTION WORD PROBLEMS

You can use ratios and proportions to solve problems.

EXAMPLE: A stick one meter long is held perpendicular to the ground and casts a shadow 0.4 meters long. At the same time, an electrical tower casts a shadow 112 meters long. Use ratio and proportion to find the height of the tower.

Step 1: Set up a proportion using the numbers in the problem. Put the shadow lengths on one side of the equation and put the heights on the other side. The 1 meter height is paired with the 0.4 meter length so let them both be top numbers. Let the unknown height be x.

$$\begin{array}{cc} \text{shadow} & \text{object} \\ \text{length} & \text{height} \end{array}$$

$$\frac{0.4}{112} = \frac{1}{x}$$

Step 2: Solve the proportion as you did on the previous page. $112 \times 1 = 112$
$112 \div 0.4 = 280$ **Answer:** The tower height is 280 meters.

Use ratio and proportion to solve the following problems.

1. Rudolph can mow a lawn that measures 1000 square feet in 2 hours. At that rate, how long would it take him to mow a lawn 3500 square feet?

2. Faye wants to know how tall her school building is. On a sunny day, she measures the shadow of the building to be 6 feet. At the same time she measures the shadow cast by a 5 foot statue to be 2 feet. How tall is her school building?

3. Out of every 5 students surveyed, 2 listen to country music. At that rate, how many students in a school of 800 listen to country music?

4. Kosco, a Labrador Retriever, had a litter of 8 puppies. Four were black. At that rate, how many would be black in a litter of 10 puppies?

5. According to the instructions on a bag of fertilizer, 5 pounds of fertilizer are needed for every 100 square feet of lawn. How many square feet will a 25 pound bag cover?

6. A race car can travel 2 laps in 5 minutes. How long will it take the race car to complete 100 laps at that rate?

7. If it takes 7 cups of flour to make 4 loaves of bread, how many loaves of bread can you make from 35 cups of flour?

8. If 3 pounds of jelly beans cost $6.30, how much would 2 pounds cost?

9. For the first 4 home football games, the concession stand sold a total of 600 hotdogs. If that ratio stays constant, how many hotdogs will sell for all 10 home games?

MAPS AND SCALE DRAWINGS

EXAMPLE 1: On a map drawn to scale, 5 cm represents 30 kilometers. A line segment connecting two cities is 7 cm long. What distance does this line segment represent?

Step 1: Set up a proportion using the numbers in the problem. Keep centimeters on one side of the equation and kilometers on the other. The 5 cm is paired with the 30 kilometers, so let them both be top numbers. Let the unknown distance be x.

$$\begin{array}{cc} \text{cm} & \text{km} \\ \dfrac{5}{7} & = \dfrac{30}{x} \end{array}$$

Step 2: Solve the proportion as you have previously. $7 \times 30 = 210$
$210 \div 5 = 42$ **Answer:** 7 cm represents 42 km.

Sometimes the answer to a scale drawing problem will be a fraction or mixed number.

EXAMPLE 2: On a scale drawing, 2 inches represents 30 feet. How many inches long is a line segment that represents 5 feet?

Step 1: Set up the proportion as you did above.

$$\begin{array}{cc} \text{inches} & \text{feet} \\ \dfrac{2}{x} & = \dfrac{30}{5} \end{array}$$

Step 2: **First multiply the two numbers that are diagonal from each other. Then divide by the other number.**

$2 \times 5 = 10$ $10 \div 30$ is less than 1, so express the answer as a fraction and reduce.

$10 \div 30 = \dfrac{10}{30} = \dfrac{1}{3}$ inch.

Set up proportions for each of the following problems and solve.

1. If 2 inches represents 50 miles on a scale drawing, how long would a line segment be that represents 25 miles? _____

2. On a scale drawing, 2 cm represents 15 km. A line segment on the drawing is 3 cm long. What distance does this line segment represent? _____

3. On a map drawn to scale, 5 cm represents 250 km. How many kilometers are represented by a line 6 cm long? _____

4. If 2 inches represents 80 miles on a scale drawing, how long would a line segment be that represents 280 miles? _____

5. On a map drawn to scale, 5 cm represents 200 km. How long would a line segment be that represents 260 km? _____

6. On a scale drawing of a house plan, one inch represents 5 feet. How many feet wide is the bathroom if the width on the drawing is 3 inches? _____

USING A SCALE TO FIND DISTANCES

By using a **map scale**, you can determine the distance between two places in the real world. The **map scale** shows distances in both miles and kilometers. You will need your ruler to do these exercises. On the scale below, you will notice that 1 inch = 800 miles. To find the distance between Calgary and Ottawa, measure with a ruler between the two cities. You will find it measures about $2\frac{1}{2}$ inches. From the scale, you know 1 inch = 800 miles. Use multiplication to find the distance in miles. $2.5 \times 800 = 2,000$. The cities are about 2,000 miles apart.

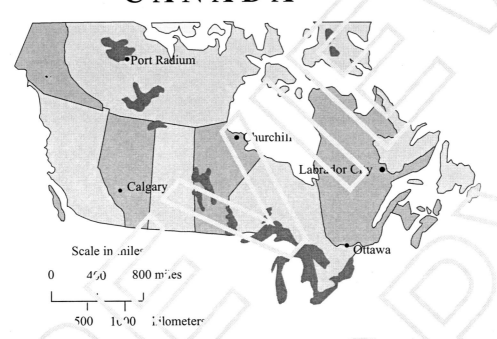

Find these distances in miles.

1. Calgary to Churchill _____
2. Churchill to Ottawa _____
3. Port Radium to Churchill _____
4. Port Radium to Ottawa _____
5. Labrador City to Ottawa _____
6. Calgary to Labrador City _____

Find these distances in kilometers.

7. Churchill to Labrador City _____
8. Ottawa to Port Radium _____
9. Port Radium to Calgary _____
10. Churchill to Ottawa _____
11. Calgary to Churchill _____
12. Calgary to Ottawa _____

USING A SCALE ON A BLUEPRINT

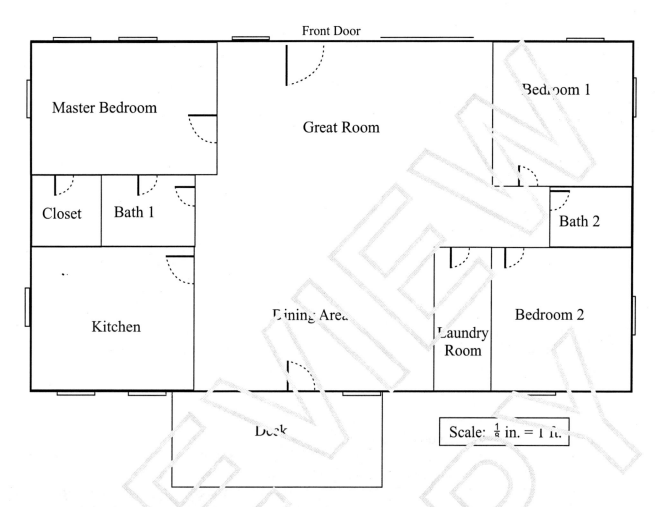

Use a ruler to find the measurements of the rooms on the blueprint above. Convert to feet using the scale. The first problem is done for you.

	long wall		short wall	
	ruler measurement	room measurement	ruler measurement	room measurement
1. Kitchen	$1\frac{3}{4}$ in.	14 ft.	$1\frac{1}{2}$ in.	12 ft.
2. Deck				
3. Closet				
4. Bedroom 1				
5. Bedroom 2				
6. Master Bedroom				
7. Bath 1				
8. Bath 2				

CHAPTER REVIEW

1. Out of 100 coins, 45 are in mint condition. What is the ratio of mint condition coins to the total number of coins?

2. What is the probability that the spinner will stop on a shaded wedge or an odd number?

3. Twenty out of the total 235 seniors graduated with honors. What is the ratio of seniors graduating with honors to the total number of seniors?

4. Fluffy's cat treat box contains 6 chicken-flavored treats, 5 beef-flavored treats, and 7 fish-flavored treats. If Fluffy's owner reaches in the box without looking, what is the probability that Fluffy will get a chicken-flavored treat?

5. Shondra used 6 ounces of chocolate chips to make two dozen cookies. At that rate, how many ounces of chocolate chips would she need to make seven dozen cookies?

6. When Rick measures the shadow of a yard stick, it is 5 inches. At the same time, the shadow of the tree he would like to chop down is 45 inches. How tall is the tree in yards?

Solve the following proportions:

7. $\frac{8}{x} = \frac{1}{2}$

8. $\frac{2}{5} = \frac{x}{10}$

9. $\frac{x}{6} = \frac{3}{9}$

10. $\frac{4}{9} = \frac{8}{x}$

11. On a scale drawing of a house floor plan, 1 inch represents 2 feet. The length of the kitchen measures 5 inches on the floor plan. How many feet does that represent?

12. If 4 inches represents 8 feet on a scale drawing, how many feet does 6 inches represent?

13. On a scale drawing, 2 cm represents 100 miles. If a line segment between two points measured 5 cm, how many miles would it represent?

14. On a map scale, 2 centimeters represents 5 km. If two towns on the map are 20 km apart, how long would the line segment be between the two towns on the map?

15. If 3 inches represents 10 feet on a scale drawing, how long will a line segment be that represents 15 feet?

CHAPTER TEST

1. In April, A+ Accountants averaged $450,000 profit. For the entire year, they averaged $750,000 profit. What is the ratio of April profit to total profit?

   A. $\frac{\$3.00}{\$5.00}$
   B. $\frac{\$3.00}{\$8.00}$
   C. $\frac{\$5.00}{\$3.00}$
   D. $\frac{\$8.00}{\$3.00}$

2. John runs 35 miles every week. At this rate, how many miles does he run in 10 days?

   A. $24\frac{1}{2}$ miles
   B. 50 miles
   C. 175 miles
   D. 350 miles

3. Fifty chickens can lay 310 eggs per week. At that rate, how many eggs can 70 chickens lay per week?

   A. 510
   B. 222
   C. 434
   D. 450

4. What is the probability that the spinner will land on a shaded section or the number 4?

   A. $\frac{1}{6}$
   B. $\frac{1}{3}$
   C. $\frac{1}{2}$
   D. $\frac{2}{3}$

5. On a blueprint of a house, 2 inches represents 3 feet. How many feet wide is a bathroom that measures 9 inches?

   A. 6 feet
   B. 9 feet
   C. 13.5 feet
   D. 18 feet

6. At the school store, 5 pens sell for $1.25. Which proportion below will help you find the cost of 12 pens?

   A. $\frac{5}{12} = \frac{\$1.25}{x}$
   B. $\frac{5}{12} = \frac{x}{\$1.25}$
   C. $\frac{5}{x} = \frac{12}{\$1.25}$
   D. $\frac{x}{5} = \frac{12}{\$1.25}$

7. On a map drawn to scale, 2 centimeters represents 300 kilometers. How long would a line measure between two cities that are 500 kilometers apart?

   A. $1\frac{1}{5}$ cm
   B. $3\frac{1}{3}$ cm
   C. 5 cm
   D. $\frac{2}{5}$ cm

8. There are 20 male and 35 female students taking band this year at Washington Middle School. What is the ratio of female students to male students taking band this year.

   A. $\frac{4}{7}$
   B. $\frac{7}{4}$
   C. $\frac{7}{11}$
   D. $\frac{11}{7}$

9. Twenty-six cards each having a different letter of the alphabet were placed face down on a table. Tyrone picked a card at random. What is the probability that he chose a vowel (A, E, I, O, or U)?

   A. $\frac{1}{26}$
   B. $\frac{5}{26}$
   C. $\frac{21}{26}$
   D. $\frac{1}{5}$

10. Rami has an aquarium with 3 black goldfish and 4 orange goldfish. He purchased 2 more black goldfish to add to his aquarium. What is the new ratio of black goldfish to total goldfish?

    A. $\frac{2}{9}$
    B. $\frac{5}{4}$
    C. $\frac{4}{5}$
    D. $\frac{5}{9}$

11. On a blueprint of a house, the scale is 0.25 inches equals 2 feet. How wide is the kitchen if it measures 1.5 inches on the blueprint?

    A. 0.03 feet
    B. 3 feet
    C. 12 feet
    D. 15 feet

12. Aunt Bess uses 3 cups of oatmeal to bake 6 dozen oatmeal cookies. How many cups of oatmeal would she need to bake 15 dozen cookies?

    A. 1.2
    B. 7.5
    C. 18
    D. 30

13. On a map, 2 centimeters represents 150 kilometers. If a line between two cities measures 5 centimeters, how many kilometers apart are they?

    A. 100 kilometers
    B. 250 kilometers
    C. 375 kilometers
    D. 750 kilometers

14. The spinner below stopped on the number 5 on the first spin. What is the probability that it will not stop on the number 5 on the second spin?

    A. $\frac{1}{3}$
    B. $\frac{1}{5}$
    C. $\frac{1}{6}$
    D. $\frac{5}{6}$

15. Rogers' Ranch has 30 chickens, 14 cows, and 3 bulls. What is the ratio of bulls to cows?

    A. $\frac{3}{5}$
    B. $\frac{1}{18}$
    C. $\frac{1}{9}$
    D. $\frac{1}{28}$

16. A pine tree casts a shadow 9 feet long. At the same time, a rod measuring 4 feet casts a shadow 1.5 feet long. How tall is the pine tree?

    A. 3.375 feet
    B. 13.5 feet
    C. 24 feet
    D. 54 feet

Chapter 6

PERCENTS

CHANGING PERCENTS TO DECIMALS AND DECIMALS TO PERCENTS

Change the following percents to **decimal** numbers.

Directions: Move the **decimal** point two places to the left and drop the **percent** sign. If there is no decimal point written, it is after the number and before the percent sign. Sometimes you will need to add a "0". (See 5% below.)

EXAMPLES: 14% = 0.14   5% = 0.05   100% = 1   103% = 1.03
(decimal point) ↑

Change the following percents to decimal numbers.

1. 18% = _____
2. 23% = _____
3. 9% = _____
4. 63% = _____
5. 4% = _____
6. 45% = _____
7. 210% = _____
8. 115% = _____
9. 2% = _____
10. 55% = _____
11. 80% = _____
12. 17% = _____
13. 66% = _____
14. 13% = _____
15. 5% = _____
16. 25% = _____
17. 410% = _____
18. 1% = _____
19. 50% = _____
20. 30% = _____
21. 107% = _____

Change the following decimal numbers to percents.

Directions: To change a **decimal** number to a **percent**, move the **decimal** point two places to the right, and add a **percent** sign. You may need to add a "0". (See 0.8 below.)

EXAMPLES: 0.62 = 62%   0.07 = 7%   0.8 = 80%   0.166 = 16.6%   1.54 = 154%

Change the following decimal numbers to percents.

22. 0.15 = _____
23. 0.062 = _____
24. 1.53 = _____
25. 0.22 = _____
26. 0.35 = _____
27. 0.375 = _____
28. 0.648 = _____
29. 0.044 = _____
30. 0.58 = _____
31. 0.86 = _____
32. 0.29 = _____
33. 0.06 = _____
34. 0.48 = _____
35. 3.089 = _____
36. 0.042 = _____
37. 0.375 = _____
38. 5.09 = _____
39. 0.75 = _____
40. 0.3 = _____
41. 2.9 = _____
42. 0.6 = _____
43. 0.122 = _____
44. 0.575 = _____
45. 0.478 = _____

CHANGING PERCENTS TO FRACTIONS AND FRACTIONS TO PERCENTS

EXAMPLE: Change 15% to a fraction.

Step 1: Copy the number without the percent sign. 15 is the top number of the fraction.

Step 2: The bottom number of the fraction is 100.

$$15\% = \frac{15}{100}$$

Step 3: Reduce the fraction. $\frac{15}{100} = \frac{3}{20}$

Change the following percents to fractions and reduce.

1. 50%
2. 13%
3. 22%
4. 95%
5. 52%
6. 63%
7. 75%
8. 91%
9. 18%
10. 3%
11. 25%
12. 5%
13. 16%
14. 1%
15. 79%
16. 40%
17. 95%
18. 30%
19. 15%
20. 84%

EXAMPLE: Change $\frac{7}{8}$ to a percent.

Step 1: Divide 7 by 8. Add as many 0's as necessary.

```
      .875
   8)7.000
    -6 4
      60
     -56
      40
     -40
       0
```

Step 2: Change the decimal answer, .875 to a percent by moving the decimal point 2 places to the right.

$$\frac{7}{8} = .875 = 87.5\%$$

Change the following fractions to percents.

1. $\frac{1}{5}$
2. $\frac{5}{8}$
3. $\frac{7}{16}$
4. $\frac{3}{8}$
5. $\frac{3}{16}$
6. $\frac{19}{100}$
7. $\frac{1}{10}$
8. $\frac{4}{5}$
9. $\frac{15}{16}$
10. $\frac{12}{25}$
11. $\frac{1}{8}$
12. $\frac{5}{16}$
13. $\frac{1}{10}$
14. $\frac{1}{4}$
15. $\frac{4}{100}$
16. $\frac{3}{4}$
17. $\frac{2}{5}$
18. $\frac{16}{25}$

CHANGING PERCENTS TO MIXED NUMBERS AND MIXED NUMBERS TO PERCENTS

EXAMPLE: Change 218% to a fraction.

Step 1: Copy the number without the percent sign. 218 is the top number of the fraction.

Step 2: The bottom number of the fraction is 100.

$$218\% = \frac{218}{100}$$

Step 3: Reduce the fraction and convert to a mixed number. $\frac{218}{100} = \frac{109}{50} = 2\frac{9}{50}$

Change the following percents to mixed numbers.

1. 150%
2. 113%
3. 222%
4. 395%
5. 252%
6. 163%
7. 275%
8. 191%
9. 108%
10. 453%
11. 205%
12. 405%
13. 516%
14. 161%
15. 179%
16. 340%
17. 199%
18. 300%
19. 125%
20. 384%

EXAMPLE: Change $5\frac{3}{8}$ to a percent.

Step 1: Divide 3 by 8. Add as many 0's as necessary.

```
    .375
8)3.000
 -2 4
    60
   -56
    40
   -40
     0
```

Step 2: So, $5\frac{3}{8} = 5.375$. Change the decimal answer to a percent by moving the decimal point 2 places to the right.

$$5\frac{3}{8} = 5.375 = 537.5\%$$

Change the following mixed numbers to percents.

1. $5\frac{1}{2}$
2. $8\frac{3}{4}$
3. $1\frac{5}{8}$
4. $3\frac{1}{4}$
5. $4\frac{7}{8}$
6. $2\frac{3}{100}$
7. $1\frac{2}{10}$
8. $6\frac{1}{5}$
9. $4\frac{7}{10}$
10. $2\frac{13}{25}$
11. $1\frac{1}{8}$
12. $2\frac{5}{16}$
13. $1\frac{3}{25}$
14. $3\frac{2}{5}$
15. $5\frac{17}{100}$
16. $4\frac{4}{5}$
17. $1\frac{1}{16}$
18. $2\frac{17}{100}$

REPRESENTING RATIONAL NUMBERS GRAPHICALLY

You now know how to convert fractions to decimals, decimals to fractions, fractions to percentages, percentages to fractions, decimals to percentages, and percentages to decimals. Study the examples below to understand how fractions, decimals, and percentages can be expressed graphically.

EXAMPLES:

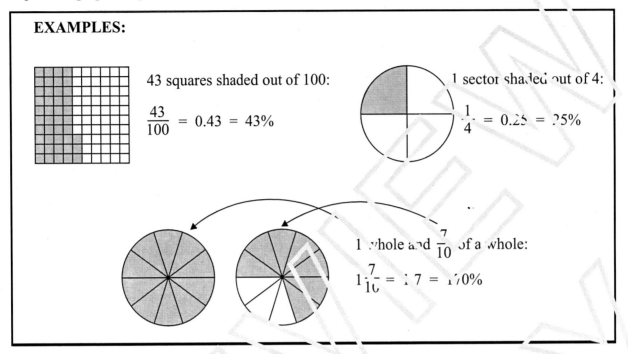

43 squares shaded out of 100:
$\frac{43}{100} = 0.43 = 43\%$

1 sector shaded out of 4:
$\frac{1}{4} = 0.25 = 25\%$

1 whole and $\frac{7}{10}$ of a whole:
$1\frac{7}{10} = 1.7 = 170\%$

Fill in the missing information in the chart below. Shade in the graphic for the problems that are not done for you. Reduce all fractions to lowest terms.

Graphic	Fraction	Decimal	Percent	Graphic	Fraction	Decimal	Percent
1.				5.			125%
2.		0.92		6.			
3.				7.			
4.	$\frac{4}{5}$			8.			

96 Copyright © American Book Company

FINDING THE PERCENT OF THE TOTAL

EXAMPLE: There were 75 customers at Billy's gas station this morning. Thirty-two percent used a credit card to make their purchase. How many customers used credit cards this morning at Billy's?

Step 1: Change 32% to a decimal.   .32

Step 2: Multiply by the total number mentioned.

```
   .32
  × 75
  ----
   160
   224
  ----
  24.00
```

24 customers used credit cards.

Find the percent of the total in the problems below.

1. Eighty-five percent of Mrs. Coomer's math class passed her final exam. There were 40 students in her class. How many passed?

2. Fifteen percent of a bag of chocolate candies have a red coating on them. How many red pieces are in a bag of 60 candies?

3. Sixty-eight percent of Valley Creek School students attended this year's homecoming dance. There are 675 students. How many attended the dance?

4. Out of the 4,500 people who attended the rock concert, forty-six percent purchased a t-shirt. How many people bought t-shirts?

5. Nina sold ninety-five percent of her 500 cookies at the bake sale. How many cookies did she sell?

6. Twelve percent of yesterday's customers purchased premium grade gasoline from GasCo. If GasCo had 200 customers, how many purchased premium grade gasoline?

7. The Candy Shack sold 138 pounds of candy on Tuesday. Fifty-two percent of the candy was jelly beans. How many pounds of jelly beans were sold Tuesday?

8. A fund-raiser at the school raised $617.50. Ninety-four percent went to local charities. How much money went to charities?

9. Out of the company's $6.5 million profit, eight percent will be paid to shareholders as dividends. How much will be paid out in dividends?

10. Ted's Toys sold seventy-five percent of its stock of stuffed bean animals on Saturday. If Ted's Toys had 620 originally in stock, how many were sold on Saturday?

COMMISSIONS

Commission: In many businesses, sales people are paid on **commission** which is a percent of the total sales they make.

EXAMPLE: Derek made a 4% commission on an $8,000 pickup truck he sold. What was the dollar amount of his commission?

Step 1 Change the percent commission to a decimal.   4% = .04

Step 2 Multiply the percent commission by the total sale.

```
        TOTAL SALE              $ 8,000
   ×    RATE OF COMMISSION      ×   .04
        COMMISSION              $ 320.00
```

Solve each of the following problems.

1. Mabel is a real estate agent who gets a 6% commission when she sells a house. How much will make on the sale of a $225,000 house?

2. Dan makes a 12% commission on the men's clothes that he sells. Last week his sales totaled $1860. How much did he earn on commission?

3. Micah earns 2% commission on the life insurance policies he sells. How much will he earn on a $80,000 policy?

4. Lane sells skin care products for a 35% commission. Last month, she sold $500.00 worth. How much was her commission?

5. Bailey earned 19% commission on her yearly sales of $158,500. How much commission did she earn for the year?

6. Carter sells vacuum cleaners for 8% commission. How much will he make on each $690 vacuum he sells?

7. Kent sells encyclopedias for $1540 per set. He earns 15% commission on each set he sells. How much does he make per set?

8. Leslie earns 10% commission on airline tickets she sells as a travel agent. Last week she sold $9540 worth of tickets. How much was her commission?

9. Pam earns 25% commission as a salesperson for her company. If she sells $1570 worth of a product, how much is her commission?

10. Caleb earns 5% commission as a car salesman. How much will he make if he sells a car for $15,590?

FINDING THE AMOUNT OF DISCOUNT

Sale prices are sometimes marked 30% off, or better yet, 50% off. A 30% DISCOUNT means you will pay 30% less than the original price. How much money you save is also known as the amount of the DISCOUNT. Read the **EXAMPLE** below to learn to figure the amount of discount.

EXAMPLE: A $179.00 chair is on sale for 30% off. How much can I save if I buy it now?

Step 1 Change 30% to a decimal. 30% = .30

Step 2 Multiply the original price by the discount.

```
ORIGINAL PRICE      $179.00
×    % DISCOUNT     ×   .30
       SAVINGS      $ 53.70
```

Practice finding the amount of the discount. Round off answers to the nearest penny.

1. Tubby Tires is offering a 25% discount on tires purchased on Tuesday. How much can you save if you buy tires on Tuesday regularly priced at $225.00 any other day of the week? _____

2. The regular price for a garden rake is $10.97 at Sly's Super Store. This week, Sly is offering a 30% discount. How much is the discount on the rake? _____

3. Christine bought a sweater regularly priced at $26.80 with a coupon for an additional 20% off. How much did she save? _____

4. The software that Marge needs for her computer is priced at $69.85. If she waits until a store offers it at 20% off, how much will she save? _____

5. Ty purchased jeans that were priced $23.97. He received a 15% employee discount. How much did he save? _____

6. The Bakery Company offers a 60% discount on all bread made the day before. How much can you save on a $2.40 loaf made today if you wait until tomorrow to buy it? _____

7. A furniture store advertises a 40% off liquidation sale on all items. How much would the discount be on a $2520 dining room set? _____

8. Becky bought a $4.00 nail polish on sale for 30% off. What was the dollar amount of the discount? _____

9. How much is the discount on a $350 racing bike marked 15% off? _____

10. Raymond receives a 2% discount from his credit card company on all purchases made with the credit card. What is his discount on $1575.50 worth of purchases? _____

FINDING THE DISCOUNTED SALE PRICE

To find the discounted sale price, you must go one step further than shown on the previous page. Read the **EXAMPLE** below to learn how to figure **discount** prices.

EXAMPLE: A $74.00 chair is on sale for 25% off. How much can I save if I buy it now?

Step 1 Change 25% to a decimal. 25% = .25

Step 2 Multiply the original price by the discount.

ORIGINAL PRICE	$74.00
× % DISCOUNT	× .25
SAVINGS	$18.50

Step 3 Subtract the savings amount from the original price to find the sale price.

ORIGINAL PRICE	$74.00
− SAVINGS	− 18.50
SALE PRICE	$55.50

Figure the sale price of the items below. The first one is done for you.

ITEM	PRICE	% OFF	MULTIPLY	SUBTRACT	SALE PRICE
1. pen	$1.50	20%	1.50 × .2 = $0.30	1.50 − 0.30 = 1.20	$1.20
2. recliner	$325	25%			
3. juicer	$55	15%			
4. blanket	$14	10%			
5. earrings	$2.40	20%			
6. figurine	$8	15%			
7. boots	$159	35%			
8. calculator	$80	30%			
9. candle	$6.20	50%			
10. camera	$445	20%			
11. VCR	$235	25%			
12. video game	$25	10%			

SALES TAX

EXAMPLE: The total price of a sofa is $560.00 + 6% sales tax. How much is the sales tax? What is the total cost?

Step 1 You will need to change 6% to a decimal. 6% = .06

Step 2 Simply multiply the cost, $560, by the tax rate, 6%. 560 × .06 = 33.6
The answer will be $33.60. (You need to add a 0 to the answer. When dealing with money, there needs to be two places after the decimal point).

```
     COST            $560
  × 6% TAX           × .06
  SALES TAX         $33.60
```

Step 3 Add the sales tax amount, $33.60 to the cost of the item sold, $560. This is the total cost.

```
     COST           $560.00
  SALES TAX       + 33.60
  TOTAL COST       $593.60
```

NOTE: When the answer to the question involves money, you always need to round off the answer to the nearest hundredth (2 places after the decimal point). Sometimes you will need to add a zero.

Figure the total costs in the problems below. The first one is done for you.

ITEM	PRICE	% SALES TAX	MULTIPLY	ADD PRICE PLUS TAX	TOTAL
1. jeans	$42	7%	$42 × 0.07 = $2.94	42 + 2.94 = 44.94	$44.94
2. truck	$17,495	6%			
3. film	$5.89	8%			
4. t-shirt	$12	5%			
5. football	$36.40	4%			
6. soda	$1.78	5%			
7. 4 tires	$105.80	10%			
8. clock	$18	6%			
9. burger	$2.34	5%			
10. software	$89.95	8%			

FINDING THE PERCENT

EXAMPLE: 15 is what percent of 60?

Step 1 To solve these problems, simply divide the smaller amount by the larger amount. You will need to add a decimal point and two 0's.

$$\begin{array}{r} .25 \\ 60\overline{)15.00} \\ -12\ 0 \\ \hline 3\ 00 \\ -3\ 00 \\ \hline 0 \end{array}$$

Step 2 Change the answer, .25, to a percent by moving the decimal point two places to the right.
.25 = 25%   15 is 25% of 60.

Remember: To change a decimal to a percent, you will sometimes have to add a zero when moving the decimal point two places to the right.

Find the following percents.

1. What percent of 50 is 16?

2. 20 is what percent of 80?

3. 9 is what percent of 100?

4. 19 is what percent of 95?

5. What percent of 200 is 25?

6. What percent of 116 is 29?

7. What percent is 18 of 90?

8. Tomika invests $36 of her $240 paycheck in a retirement account. What percent of her pay is she investing?

9. Ray sold a house for $115,000, and his commission was $9,200. What percent commission did he make?

10. Peter was making $16.00 per hour. After one year, he received a $2.00 per hour raise. What percent raise did he get?

11. Calvin budgets $235 per month for food. If his salary is $940 per month, what percent of his salary does he budget for food?

12. Katie earned $45 on commission for her sales totaling $225. What percent was her commission?

UNDERSTANDING SIMPLE INTEREST *

I = PRT is a formula to figure out the **cost of borrowing money** or the **amount you earn** when you **put money in a savings account**. When you want to buy a used truck or car, you go to the bank to borrow the $7,000 you need. The bank will charge you interest on the $7,000. If the simple interest rate is 9% for four years, you can figure the cost of the interest with this formula.

First, you need to understand these terms:

    I = Interest = The amount charged by the bank or other lender
    P = Principal = The amount you borrow
    R = Rate = The interest rate the bank is charging you
    T = Time = How many years you will take to pay off the loan

EXAMPLE:

In the problem above: **I = PRT** This means the **interest** equals the **principal** times the **rate** times the **time** in **years**.

$$I = \$7,000 \times 9\% \times 4 \text{ years}$$
$$I = \$7,000 \times .09 \times 4$$
$$I = \$2,520$$

Use the formula I = PRT to work the following problems:

1. Craig borrowed $1,800 from his parents to buy a stereo. His parents charged him 3% simple interest for 2 years. How much interest did he pay his parents? _____

2. Raul invested $5,000 in a savings account that earned 2% simple interest. If he kept the money in the account for 5 years, how much interest did he earn? _____

3. Bridgette borrowed $11,000 to buy a car. The bank charged 12% simple interest for 7 years. How much interest did she pay the bank? _____

4. A tax accountant invested $25,000 in a money market account for 3 years. The account earned 5% simple interest. How much interest did the accountant make on his investment? _____

5. Linda Kay started a savings account for her nephew with $2,000. The account earned 5% simple interest. How much interest did the account accumulate in 3 years? _____

6. Renada bought a living room set on credit. The set sold for $2,300 and the store charged her 9% simple interest for one year. How much interest did she pay? _____

7. Duane took out a $3,500 loan at 8% simple interest for 3 years. How much interest did he pay for borrowing the $3,500? _____

* Simple interest is not commonly used by banks and other lending institutions. Compound interest is more commonly used, but its calculations are more complicated and beyond the scope of the material presented in this text.

BUYING ON CREDIT

Many stores will allow you to choose between paying cash or buying on credit. When you buy on credit, you make a down payment and monthly payments until the item is paid for. You will always pay more for an item by buying on credit.

EXAMPLE: Darlene saw a living room set advertised for $899.00 or $100.00 down and $60.00 per month for 18 months. How much can she save by paying cash?

Step 1: Multiply the number of months by the monthly payments. $60.00 × 18 = $1080

Step 2: Add the down payment to the total cost of the payments. $1,080 + $100 = $1,180

Step 3: Subtract the cash price from the total cost of paying on credit. $1180 − $899 = $281

Answer: Darlene can save $281 by paying cash for the living room set.

For each problem below, determine how much you can save if you pay cash.

1.

   Notebook Computer
   $2595 cash
   or
   $300 down and $75 per mo. for 36 mo.

2.

   SALE
   Big Al's Used Jeeps
   $6000 cash
   OR
   only $500 down &
   $250 per mo. for 30 months

3.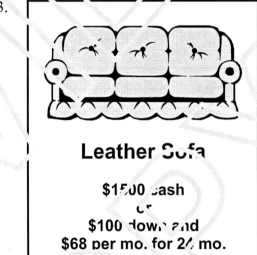

   Leather Sofa
   $1500 cash
   or
   $100 down and
   $68 per mo. for 24 mo.

4.

   JET SKI SALE
   $3395 cash
   or finance for
   $250 down &
   $145 per mo. for 24 mo.

CHAPTER REVIEW

Change the following percents to decimals.

1. 45% _____
2. 219% _____
3. 22% _____
4. 1.25% _____

Change the following decimals to percents.

5. 0.52 _____
6. 0.64 _____
7. 1.09 _____
8. 0.625 _____

Change the following percents to fractions.

9. 25% _____
10. 3% _____
11. 68% _____
12. 102% _____

Change the following fractions to percents.

13. $\frac{9}{10}$ _____
14. $\frac{5}{16}$ _____
15. $\frac{1}{8}$ _____
16. $\frac{1}{4}$ _____

17. What is 1.65 written as a percent?

_____

18. What is $2\frac{1}{2}$ written as a percent?

_____

19. Change 5.65 to a percent.

_____

Fill in the equivalent numbers represented by the shaded area.

20. fraction _____
21. decimal _____
22. percent _____

Fill in the equivalent numbers represented by the shaded area.

23. fraction _____
24. decimal _____
25. percent _____

26. Uncle Howard left his only niece 56% of his assets according to his will. If his assets totaled $564,000 when he died, how much did his niece inherit?

_____

27. Celeste makes 6% commission on her sales. If her sales for a week total $4580, what is her commission?

_____

28. Peeler's Jewelry is offering a 30% off sale on all bracelets. How much will you save if you buy a $45.00 bracelet during the sale?

_____

29. How much would an employee pay for a $724.00 stereo if the employee got a 15% discount?

_____

30. Misha bought a CD for $14.95. If sales tax was 7%, how much did she pay total?

_____

31. The Pep band made $640 during a fund-raiser. The band spent $400 of the money on new uniforms. What percent of the total did they spend on uniforms?

_____

32. Linda took out a simple interest loan for $7,000 at 11% interest for 5 years. How much interest did she have to pay back?

_____

33. McMartin's is offering a deal on fitness club memberships. You can pay $999 up front for a 3 year membership, or pay $200 down and $20 per month for 36 months. How much would you save by paying up front?

_____

34. Patton, Patton and Clark, a law firm, won a malpractice law suit for $4,500,000. Sixty-eight percent went to the law firm. How much did the law firm make?

_____

35. Jeneane earned $340.20 commission by selling $5670 worth of products. What percent commission did she earn?

_____

36. Tara put $500 in a savings account that earned 3% simple interest. How much interest did she make after 5 years?

_____

37. Clay bought a pair of basketball shoes for $79.99 plus 5% sales tax. What was his total cost for the shoes?

_____

38. Marsha earns a 12% commission on all jewelry sales. How much commission will she make if she sells a $564.00 necklace?

_____

39. The following advertisement is in the newspaper:

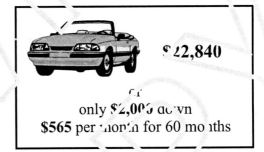

$22,840

or

only $2,000 down
$565 per month for 60 months

How much more would you pay to buy the car on credit?

_____

40. A department store is selling all swimsuits for 40% off in August. How much would you pay for a swimsuit that is normally priced at $35.80?

_____

41. How much can Jane save by waiting to buy a $34.00 sweater that will go on sale for 20% off on Saturday?

_____

CHAPTER TEST

1. 36% written as a fraction is

   A. $\frac{9}{25}$

   B. $\frac{1}{3}$

   C. $\frac{1}{2}$

   D. $\frac{1}{36}$

2. $\frac{3}{8}$ written as a percent is

   A. 37.5%
   B. 38%
   C. 3.75%
   D. 30%

3. $4\frac{1}{4}$ written as a percent is

   A. 4.14%
   B. 41.4%
   C. 42.5%
   D. 425%

4. 7.02 is the same as

   A. 7.02%
   B. 70.2%
   C. 702%
   D. 0.0702%

5. 122% written as a fraction is

   A. $1\frac{11}{50}$

   B. $2\frac{3}{25}$

   C. $12\frac{1}{5}$

   D. 122

6. 34% written as a decimal is

   A. 0.34
   B. 3.4
   C. 34.0
   D. 340

7. The shaded area of the graph below represents what percent of the total?

   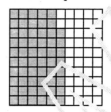

   A. 42%
   B. 58%
   C. 60%
   D. 580%

8. Tate is saving his money to buy a go-cart. The go-cart he wants costs $240. His mother agreed to chip in $60. What percent of the total cost of the go-cart is his mother contributing?

   A. 20%
   B. 25%
   C. 40%
   D. 75%

9. Mattie earns 12% commission on the perfume that she sells. How much is her commission on a $45.00 sale?

   A. $1.20
   B. $3.75
   C. $5.40
   D. $12.00

10. **Fish Palace — Going Out of Business Sale**
    All fish, aquariums, & accessories
    65% off

    According to the ad above, how much would you save by buying an aquarium that regularly sold for $54.00?

    A. $18.90
    B. $20.00
    C. $32.50
    D. $35.10

11. Paula bought a new CD player for $235.00. She paid 6% sales tax. How much did she pay total?

    A. $ 14.10
    B. $220.90
    C. $249.10
    D. $376.00

12. The winning pie-eating contestant ate 92% of the 25 pies. How many pies did the contestant eat?

    A. 20
    B. 21
    C. 22
    D. 23

13.

    Sports Unlimited
    30% off
    All Baseball Equipment

    Manny saw the above ad and decided to buy a baseball glove from Sports Unlimited during their sale. What would the discounted sale price be for a glove regularly priced at $25.50?

    A. $ 7.65
    B. $17.85
    C. $25.20
    D. $35.15

14. Mark invested $1,340 in a simple interest account at 3.5% interest. How much interest did he earn after 4 years?

    A. $ 46.90
    B. $ 187.60
    C. $ 469.00
    D. $1,876.00

15. What percent of 200 is 120?

    A. 12%
    B. 20%
    C. 55%
    D. 60%

16. Patrick sells men's shoes on commission. He earns $5.00 commission for every $100.00 in shoes that he sells. What percent commission does he earn?

    A. 5%
    B. 20%
    C. 25%
    D. 50%

17. A car dealership advertises in the paper that you can buy a new truck for $28,499 or finance the same truck for $2,000 down and $595 per month for 60 months. How much would you save by paying up front?

    A. $ 4,901
    B. $ 7,801
    C. $19,801
    D. $25,499

18. Out of the 144 players participating in Little League baseball, 18 were picked for the All-Star game. What percent of the total were picked for the All-Star game?

    A. 8%
    B. 12.5%
    C. 14.4%
    D. 18%

19. Jamie borrowed $800 from his parents as a simple interest loan at 5% interest. If he pays his parents back in two years, how much interest will he owe them?

    A. $ 40
    B. $ 60
    C. $ 80
    D. $160

20. Renata bought a stuffed bear priced at $14.80 and paid 5% sales tax. What was the total price of her purchase?

    A. $14.06
    B. $14.85
    C. $15.00
    D. $15.54

Chapter 7

PROBLEM-SOLVING AND CRITICAL THINKING

MISSING INFORMATION

Problems can only be solved if you are given enough information. Sometimes you are not given enough information to solve a problem.

EXAMPLE: Chuck has worked on his job for 1 year now. At the end of a year, his employer gave him a 12% raise. How much does Chuck make now?

To solve this problem, you need to know how much Chuck made when he began his job one year ago.

Each problem below does not give enough information for you to solve it. Beneath each problem, describe the information you would need to solve the problem.

1. Fourteen percent of the coated chocolate candies in Nate's bag were yellow. At that rate, how many of the candies were yellow?

   _____

2. Patrick is putting up a fence around all four sides of his back yard. The fence costs $2.25 per foot, and his yard is 150 feet wide. How much will the fence cost?

   _____

3. Staci worked 5 days last week. She made $6.75 per hour before taxes. What was her total earnings before taxes were taken out?

   _____

4. Which is a better buy: 4 oz bar of soap for 88¢ or a bath bar for $1.29?

   _____

5. Pandy bought a used car for $4,568 plus sales tax. What was the total cost of the car?

   _____

6. The Hall family ate at a restaurant, and each of their dinners cost $5.95. They left a 15% tip. What was the total amount of the tip?

   _____

7. If a kudzu plant grows 3 feet per day, in what month will it be 90 feet long?

   _____

8. Bethany traveled by car to her sister's house in Raleigh. She traveled at an average speed of 52 miles per hour. She arrived at 4:00 p.m. How far did she travel?

   _____

9. Terrence earns $7.50 per hour plus 5% commission on total sales over $500 per day. Today he sold $6,500 worth of merchandise. How much did he earn for the day?

_____

10. Michelle works at a department store and gets an employee's discount on all of her purchases. She wants to buy a sweater that sells for $38.00. How much will the sweater cost after her discount?

_____

11. John filled his car with 10 gallons of gas and paid for the gas with a $20 bill. How much change did he get back?

_____

12. Olivia budgets $5.00 per work day for lunch. How much does she budget for lunch each month?

_____

13. Joey worked 40 hours and was paid $356.00. His friend Pete worked 38 hours. Who was paid more per hour?

_____

14. A train trip from Columbia to Boston took $13\frac{1}{4}$ hours. How many miles apart are the two cities?

_____

15. Caleb spent 35% of his check on rent, 10% on groceries, and 18% on utilities. How much money did he have left from his check?

_____

16. The Lyons family spent $54.00 per day plus tax on lodging during their vacation. How much tax did they pay for lodging per day?

_____

17. Richard bought cologne at a 30% off sale. How much did he save buying the cologne on sale?

_____

18. The bottling machine works 7 days a week and fills 1,000 bottles per hour. How many bottles did it fill last week?

_____

19. Tyler, who works strictly on commission, brought in $25,000 worth of sales in the last 10 days. How much was his commission?

_____

20. Ninety percent of the student body at Paris Middle School bought raffle tickets to help the basketball team buy new uniforms. The main prize was a 25 inch color TV with built-in VCR. How many students bought raffle tickets?

_____

EXACT INFORMATION

Most word problems supply exact information and ask for exact answers. The following problems are the same as those on the previous two pages with the missing information given. Find the exact solution.

1. Fourteen percent of the coated chocolate candies in Nate's bag were yellow. If there were 50 pieces in the bag, how many of the candies were yellow?

2. Patrick is putting up a fence around all four sides of his back yard. The fence costs $2.25 per foot. His yard is 150 feet wide and 200 feet long. How much will the fence cost?

3. Staci worked 5 days last week, 8 hours each day. She made $6.75 per hour before taxes. How much did she make last week before taxes were taken out?

4. Which is a better buy. 4 oz bar of soap for 85¢ or a 6 oz bath bar for $1.20?

5. Randy bought a used car for $4,568 plus 6% sales tax. What was the total cost of the car?

6. The Hall family ate at a restaurant, and each of the 4 dinners cost $5.95. They left a 15% tip. What was the total amount of the tip?

7. If a kudzu plant grows 2 feet per day, in what month will it be 90 feet long if it takes root in the middle of May?

8. Bethany traveled from South Carolina by car to her sister's house in Raleigh, North Carolina. She traveled at an average speed of 52 miles per hour. She left at 10:00 a.m. and arrived at 4:00 p.m. How far did she travel?

9. Terrence earns $7.50 per hour plus 5% commission on total sales over $500 per day. Today he sold $6,500 worth of merchandise and worked 7 hours. How much did he earn for the day?

10. Michelle works at a department store and gets a 20% employee's discount on all of her purchases. She wants to buy a sweater that sells for $38.00. How much will the sweater cost after her discount?

11. John filled his car with 10 gallons of gas priced at $1.24 per gallon. He paid for the gas with a $20 bill. How much change did he get back?

12. Olivia budgets $5.00 per work day for lunch. How much does she budget for lunches if she works 21 days this month?

13. Joey worked 40 hours and was paid $356.00. His friend Pete worked 38 hours at $8.70 per hour. Who was paid more per hour?

14. A train trip from Columbia, SC to Boston, MA took $18\frac{1}{4}$ hours. How many miles apart are the two cities if the train travels at an average speed of 60 miles per hour?

15. Caleb spent 35% of his check on rent, 10% on groceries, and 18% on utilities. How much money did he have left from his $260 check?

16. The Lyons family spent $54.00 per day plus 10% tax on lodging during their vacation. How much tax did they pay per day?

17. Richard bought cologne at a 30% off sale. The cologne was regularly priced at $44. How much did he save buying the cologne on sale?

18. The bottling machine works 7 days a week, 14 hours per day and fills 1,000 bottles per hour. How many bottles did it fill last week?

19. Tyler, who works strictly on commission, brought in $25,000 worth of sales in the last 10 days. He earns 15% commission on his sales. How much was his commission?

20. Ninety percent of the total 540 students at Parks High School bought raffle tickets to help the basketball team buy new uniforms. How many students bought raffle tickets?

EXTRA INFORMATION

In each of the following problems, there is extra information given. **Look closely at the question,** and use only the information you need to answer it.

EXAMPLE: Gary was making $6.50 per hour. His boss gave him a 52¢ per hour raise. Gary works 40 hours per week. What percent raise did Gary receive?

Solution: To figure the percent of Gary's raise, you do **not** need to know how many hours per week Gary works. That is extra information not needed to answer the question. To figure the percent increase, simply divide the change in pay, $0.52, by the original wages, $6.50. $0.52 \div 6.50 = 0.08$

Gary received an 8% raise.

In the following questions, determine what information is needed from the problem to answer the question and solve.

1. Leah wants a new sound system that is on sale for 15% off the regular price of $420. She has already saved $325 toward the cost. What is the dollar amount of the discount?

2. Praveen bought a shirt for $24.80 and socks for $11.25. He gets $10.00 per week for his allowance. He paid $2.76 sales tax. What was his change from three $20 bills?

3. Marty worked 38 hours this week, and he earned $8.40 per hour. His taxes and insurance deductions amount to 34% of his gross pay. What is his total gross pay?

4. Tamika went shopping and spent $4.80 for lunch. She wants to buy a sweater that is on sale for $\frac{1}{4}$ off the regular $56.00 price. How much will she save?

5. Nick drove an average of 52 miles per hour for 7 hours. His car gets 32 miles per gallon. How far did he travel?

6. The odometer on Melody's car read 45,920 at the beginning of her trip and 46,400 at the end of her trip. Her speed averaged 54 miles per hour, and she used 20 gallons of gasoline. How many miles per gallon did she average?

7. Eighty percent of the eighth graders attended the end of the year class picnic. There are 160 eighth graders and 54% of them ride the bus to school each day. How many students went to the class picnic?

8. Matt has $5.00 to spend on snacks. Tastee Potato Chips cost $2.57 for a one pound bag at the grocery store. T-Mart sells the same bag of chips for $1.98. How much can he save if he buys the chips at T-Mart?

9. Elaina wanted to make 10 cakes for the band bake sale. She needed $1\frac{3}{4}$ cups of flour and $2\frac{1}{4}$ cups of sugar for each cake. How many cups of flour did she need in all?

ESTIMATED SOLUTIONS

Some problems require an estimated solution. In order to have enough product to complete the job, you often need to buy more than you actually need because of the sizes the product come in. **In the following problems, be sure to round up your answer to the next whole number to find the correct solution.**

1. Endicott Publishing received an order for 550 books. Each shipping box holds 30 books. How many boxes do the packers need to ship the order?

   _____

2. Elena's 250 chickens laid 314 eggs in the last 2 days. How many egg cartons holding one dozen eggs would be needed to hold all the eggs?

   _____

3. Antoinetta's Italian restaurant uses $1\frac{1}{4}$ quarts of olive oil every day. The restaurant is open 7 days a week. For the month of September how many gallons should they order to have enough?

   _____

4. Eastmont High School is taking 516 students and 22 chaperones on a field trip. Each bus holds 44 persons. How many buses will the school need?

   _____

5. Fran volunteered to hem 11 choir robes that came in too long. Each robe is 7 feet around at the bottom. Hemming tape comes three yards to a pack. How many packs will Fran need to buy to go around all the robes?

   _____

6. Tonya is making matching vests for the children's choir. Each vest has 5 buttons on it, and there are 23 children in the choir. The button she picked comes 6 buttons to a card. How many cards of buttons does she need?

   _____

7. Tiffany is making the bread for the banquet. She needs to make 6 batches with $2\frac{1}{4}$ lb of flour in each batch. How many 10 lb bags of flour will she need to buy?

   _____

8. The homeless shelter is distributing 250 sandwiches per day to hungry guests. It takes one foot of plastic wrap to wrap each sandwich. There are 150 feet of plastic wrap per box. How many boxes will Mary need to buy to have enough plastic wrap for the week?

   _____

9. An advertising company has 15 different kinds of one-page flyers. The company needs 75 copies of each kind of flyer. How many reams of paper will the company need to produce the flyers? One ream equals 500 sheets of paper.

   _____

TWO-STEP PROBLEMS

Some problems require two steps to solve.

Read each of the following problems carefully and solve.

1. For a family picnic, Renee bought 10 pounds of hamburger meat and used $\frac{1}{4}$ pound of meat to make each hamburger patty. Renee's family ate 32 hamburgers. How many pounds of hamburger meat did she have left?

2. Vic sold 45 raffle tickets. His brother sold twice as many. How many tickets did they sell together?

3. Erin earns $2,200 per month. Her deductions amount to 28% of her paycheck. How much does she take home each month?

4. Matheson Middle School Band is selling t-shirts to raise money for new uniforms. They need to raise $1260. They are selling t-shirts for $12 each. There is a $6 profit for each shirt sold. So far, they have sold 80 t-shirts. How many more t-shirts do they need to sell to raise the $1260?

5. Alphonso was earning $1,860 per month and then got a 12% raise. How much will he make per month now?

6. Barbara and Jeff ate out for dinner. The total came to $15.00. They left a 15% tip. How much was the tip and the meal together?

7. Hillary is bicycling across Montana taking a 845 mile course. The first week, she covered 320 miles. The second week she traveled another 350 miles. How many more miles does she have to travel to complete the course?

8. Jason budgets 30% of his $1,100 income each month for food. How much money does he have to spend for everything else?

9. After Madison makes a 12% down payment on a $2,000 motorcycle, how much will she still owe?

10. Randy bought a pair of shoes for $51, a tie for $18, and a new belt for $23. If the sales tax is 8%, how much sales tax did he pay?

PATTERN PROBLEMS

Some problems follow a pattern. You must read these problems carefully and recognize the pattern.

EXAMPLE: Jason wants to swim in the ocean from the shore to the end of a pier 38 feet out. He must swim against the tide. For every 10 feet he swims forward, the tide takes him back 3 feet. How many total feet will he swim to reach the end of the pier?

Step 1: Draw a diagram or create a table to help you visualize the problem.

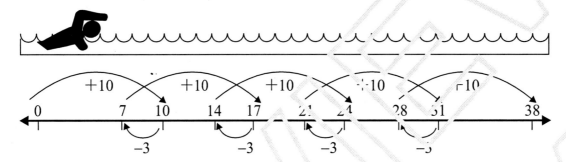

Step 2: Determine how many feet he swam by going forward 10 feet at a time. Look at the diagram above. There are 5 forward arrows that each represent 10 feet to get Jason to the 38 feet mark.
10 × 5 = 50 feet

Read each of the following questions carefully. Draw a diagram or create a table to help you visualize the pattern. Then answer the question.

1. In a strong man contest, the contestants must pull a car up an incline 13 feet long. Joe is the first contestant. With every tug, Joe pulls the car up the incline 3 feet. Before he can tug again, the car rolls down 1 foot. How many tugs will it take Joe to get the car up the incline?

2. Greta goes on a jungle safari and steps in quicksand. She sinks 25 inches. Her partner loops a rope around a tree limb directly above her so she can pull herself out. For every 5 inches she is able to pull up, she slips back down 1 inch. How many total inches must she climb up to completely free herself from the 25 inches of quicksand?

3. Randal is starting a job as a commissioned salesman. He establishes 5 clients the first week of his job. His goal is to double his clientele every week. If he can accomplish his goal, how many weeks will it take him to establish 320 clients?

4. A five-year old decides to run the opposite way on the moving sidewalk at the airport. For every 7 feet he runs forward, the sidewalk moves him 4 feet backward. How many total feet will he run to get to the end of a sidewalk that is 19 feet long?

5. A certain bacteria culture doubles its population every minute. Every 3 minutes, the entire culture decreases by half. If a culture starts with a population of 1000, what would be the population of the bacteria at 6 minutes? **Hint. The population equals 1000 at 0 minutes.**

6. Rita's mom needs to bake 5 dozen chocolate chip cookies for a bake sale. For every dozen cookies she bakes, her children eat two. How many dozen cookies will she have to bake to have 5 dozen for the bake sale?

7. A retail store plans on building 3 new stores each year. They expect that 1 store will go out of business every $2\frac{1}{2}$ years. How many years would it take to establish 26 stores?

8. Jasmine and her family sold Girl Scout cookies this spring at the mall. After every seven boxes she sold, her family bought a box to celebrate. Jasmine sold a total of 64 boxes. How many of those boxes did her family buy?

9. A pitcher plant has leaves modified as pitchers for trapping and digesting insects. An insect is lured to the edge of a 7 inch leaf and slips down 3 inches. For every 3 inches it slips down, it manages to climb up 2 inches. It then starts slipping back down 3 inches towards the digestive juices at the bottom of the leaf. How many total inches will the insect slip down until it reaches the bottom?

PATTERNS IN EXCHANGES

Some problems involve people exchanging things. Carefully read the example below.

EXAMPLE: Skip, Joe, Pete, and Don are new neighbors. The four neighbors meet together to exchange phone numbers in case of an emergency. Each neighbor writes down his number for each of the other neighbors. What is the total number of telephone numbers written down?

Method 1: Make a list or draw a diagram to figure out the total number of phone numbers written. Skip writes his number once for Joe, once for Pete, and once for Don. Joe does the same for Skip, Pete, and Don. Pete copies his number for Skip, Joe, and Don. Finally, Don writes his number for Skip, Joe, and Pete. If you count the total number of phone numbers exchanged, you should get 12.

Method 2: Each neighbor knows his own number, so each writes his number for the other 3 neighbors. So 4 neighbors write their numbers down 3 times. 4 × 3 = 12.

In this type of problem where people are exchanging a unique item, you can use the following formula:

Total exchanges = (the number of people) × (the number of people − 1)

Read the following questions carefully, and then solve each by using one of the methods above. Please note that the formula given in method 2 may not be appropriate to solve every problem.

1. At summer camp, 5 friends exchanged home addresses so that they could all keep in touch by writing. What is the total number of addresses exchanged between the 5 friends?

2. In Mrs. Kirk's homeroom class, all 20 students exchanged their yearbooks to be signed by each of their peers. How many signatures were written?

3. Three co-workers exchanged e-mail addresses with one another. How many e-mail addresses were exchanged?

4. A creative writing class held a contest for best written short story. Four friends in the class decided to write two short stories each. They then exchanged their written stories with each other to proofread before turning them in. How many stories did each friend receive to read?

5. A class of 15 students had a party. Each student brought a different kind of candy to share with the other students. Each student started with an empty bag. The students then passed their bags around to each of the other students to receive some of the candy each brought in. How many times must each student pass his or her bag to sample the candy from all the other students? How many total passes were made among all the students?

118 Copyright © American Book Company

MORE PATTERNS IN EXCHANGES

How does the number of exchanges between people change if the exchange is something like a handshake? Look at the example below where two people exchange one thing, like a handshake.

EXAMPLE: In a doubles tennis match, a team of two players competes against a team of two other players. After a match, it is customary for each of the players to shake hands at the net including a handshake between teammates. How many total handshakes are exchanged?

Long Method: Make a list or draw a diagram to figure out the total number of handshakes. If you count the handshakes represented by the diagram below, you should get 6.

Short Method: Use the same formula as you did on the previous page, but since two people are exchanging one thing, divide your answer by 2. $4 \times 3 = 12$ and $12 \div 2 = 6$.

When people are exchanging a mutual item, you can use the following formula:

$$\text{Total exchanges} = \frac{(\text{the number of people}) \times (\text{the number of people} - 1)}{2}$$

Read the following questions carefully, and then solve each by using one of the methods above. Please note that the formula given in the short method above may not be appropriate to solve every problem.

1. The last day of summer camp, 5 girls who had become good friends gave each other a hug. How many hugs were exchanged?

   _____

2. Before the coin toss at the football game, three captains from the Bruins team shook hands with each of the four captains from the Warriors team. How many handshakes were exchanged?

   _____

3. Seven basketball teams in a conference play each other twice in a season. How many total games are played in a season?

   _____

4. The teacher gave each of her 10 students a pegboard with 9 pegs. Each student was asked to connect each of the pegs on his or her pegboard with a rubber band. How many rubber bands did each student use to connect all the pegs? How many rubber bands did the entire class use?

   _____

   _____

5. Six different peaceful tribes built roads between each of their villages for easy trade. How many roads did they create if a different road connects each village?

   _____

USING DIAGRAMS TO SOLVE PROBLEMS

Problems that require logical reasoning cannot always be solved with a set formula. Sometimes, drawing diagrams can help you see the solution.

EXAMPLE: Yvette, Barbara, Patty, and Nicole agreed to meet at the movie theater around 7:00 p.m. Nicole arrived before Yvette. Barbara arrived after Yvette. Patty arrived before Barbara but after Yvette. What is the order of their arrival?

Nicole	Yvette	Patty	Barbara
1st	2nd	3rd	4th

Arrange the names in a diagram so that the order agrees with the problem.

Use a diagram to answer each of the following questions

1. Javy, Thomas, Pat, and Keith raced their bikes across the playground. Keith beat Thomas but lost to Pat and Javy. Pat beat Javy. Who won the race?

2. Jeff, Greg, Pedro, Lisa, Macy, and Kay eat lunch together at a round table. Kay wants to sit beside Pedro. Pedro wants to sit next to Lisa, Greg wants to sit next to Macy, and Jeff wants to sit beside Kay. Macy would rather not sit beside Lisa. Which two people should sit on each side of Jeff?

3. Three teams play a round-robin tournament where each team plays every other team. Team A beat Team C, Team B beat Team A. Team B beat Team C. Which team is the best?

4. Caleb, Thomas, Ginger, Alex, and Janice are in the lunch line. Thomas is behind Alex. Caleb is in front of Alex but behind Ginger and Janice. Janice is between Ginger and Caleb. Who is third in line?

5. Ray, Fleta, Paula, Joan, and Henry hold hands to make a circle. Joan is between Ray and Paula. Fleta is holding Ray's other hand. Paula is also holding Henry's hand. Who must be holding Henry's other hand?

6. The Bears, the Cavaliers, the Knights, and the Lions all competed in a track meet. One team from each school ran the 400 meter relay race. The Bears beat the Knights but lost to the Cavaliers. The Lions beat the Cavaliers. Who finished first, second, third, and fourth?

   1st _____

   2nd _____

   3rd _____

   4th _____

TRIAL AND ERROR PROBLEMS

Sometimes problems can only be solved by trial and error. You have to guess at a solution, and then check to see if it will satisfy the problem. If it does not, you must guess again until you get the right answer.

Solve the following problems by trial and error. Make a chart of your attempts so that you don't repeat the same attempt twice.

1. Becca had 5 coins consisting of one or more quarters, dimes, and nickels that totaled 75¢. How many quarters, dimes, and nickels did she have?

    quarters _____

    dimes _____

    nickels _____

2. Ryan needs to buy 42 cans of soda for a party at his house. He can get a six pack for $1.80, a box of 12 for $3.00, or a case of 24 for $4.90. What is the least amount of money Ryan must spend to purchase the 42 cans of soda?

3. Jana had 10 building blocks that were numbered 1 to 10. She took three of the blocks and added up the three numbers to get 27. Which three blocks did she pick?

4. Hank had 10 coins. He had 3 quarters, 3 dimes, and 4 nickels. He bought a candy bar for 75¢. How many different ways could he spend his coins to pay for the candy bar?

5. Refer to question 5. If Hank used 6 coins to pay for the candy bar, how many of his quarters did he spend?

6. The junior varsity basketball team needs to order 38 pairs of socks for the season. The coach can order 1 pair for $2.45, 6 pairs for $12.95, or 10 pairs for $20.95. What is the least amount of money he will need to spend to purchase exactly 38 pairs of socks?

7. Tyler has 5 quarters, 10 dimes, and 15 nickels in change. He wants to buy a notebook for $2.35 using the change that he has. If he wants to use as many of the coins as possible, how many quarters will he spend?

8. Kevin is packing up his room to move to another city. He has the following items left to pack.

    comic book collection ... 7 pounds
    track trophy ... 3 pounds
    coin collection ... 13 pounds
    soccer ball ... 1 pound
    model car ... 6 pounds

    If Kevin has a large box that will hold 25 pounds, what items should he pack in it to get the most weight without going over the box's weight limit?

CHAPTER REVIEW

Read each problem carefully and solve using the problem-solving methods you learned in this chapter. If there is not enough information, tell what information is missing.

1. East Point Middle School is taking 316 students, 12 teachers, and 10 parents on a field trip. Each bus holds 44 persons. How many busses will the school need?

2. Kyle paid for his school lunch with a $10 bill. How much change did he get back?

3. Tyrone bought shoes for $65 and a shirt for $24. He paid 8% sales tax. How much did he pay for the two items?

4. Crystal bought a jacket at a 20% off sale. How much did she save buying the jacket on sale?

5. Laura left her house at 8:30 a.m. in Charlotte, NC and drove 5 hours to see her brother in Atlanta, GA. Her trip from Charlotte to Atlanta totaled 260 miles. What was her average speed in miles per hour?

6. Nathan drove 4 hours to get home. He arrived at 10:00 p.m. How fast did he drive?

7. Eight soccer teams make up a conference. Each team plays the others three times during a season. How many games does each team play?

8. Tran worked 4 days last week. She made $6.26 per hour. Her employer deducted 24% of her pay for taxes. How much was Tran's take-home pay?

9. Lori bought a sweater originally marked $32.50 and a belt for $12.25. The sweater was on sale for 20% off. How much money did she save by buying the sweater on sale?

10. Six football teams compete in a conference. Each team plays the other teams twice a season. How many total football games are played each season?

11. Cecilia won 1,390 tickets in the arcade. A stuffed bear costs 474 tickets, a stop watch costs 1,620, and a clown wig costs 1,240 tickets. How many more tickets does Cecilia need to win to purchase the stop watch and the stuffed bear?

12. Andrea spent $1.24 for toothpaste. She had quarters, dimes, nickels, and pennies. How many dimes did she use if she used a total of 14 coins?

13. Seth left his home at 2:00 p.m. He arrived at the airport at 3:30 p.m., $1\frac{1}{2}$ hours before his plane departed. What was Seth's average speed traveling to the airport?

14. Cody spent 59¢ on a hotdog. He had quarters, dimes, nickels and pennies in his pocket. He gave the cashier 9 coins for exact payment. How many quarters did he give the cashier?

_____

15. Carlyle is filling a 15 gallon drum with water but doesn't realize that the drum has a leak in the bottom. For every 3 gallons he puts in, 1 gallon leaks out. How many gallons will he use to fill the drum up to the 15 gallon mark?

_____

16. Five friends get together for a barbeque and exchange hugs. If each friend hugs every other friend one time, how many hugs were exchanged?

_____

17. Vince, Hal, Weng and Carl raced on rollerblades down a hill. Vince beat Carl. Hal finished before Vince but after Weng. Who won the race?

_____

18. After a business luncheon, 5 businessmen exchanged their business cards with one another. How many business cards were exchanged?

_____

19. Mrs. Rhodes gave each of her 35 second grade students a Valentine's Day card. The cards that she picked out came 12 to a box. How many boxes did she have to purchase?

_____

20. A veterinarian's office has a weight scale in the lobby to weigh pets. The scale will weigh up to 250 pounds. The following dogs are in the lobby:

    Pepper ... 23 pounds
    Jack ... 75 pounds
    Trooper ... 45 pounds
    Precious ... 25 pounds
    Coco ... 120 pounds

Which dogs could you put on the scale to get as close to 250 pounds without going over? How much would they weigh?

_____

_____

21. The Knights and the Eagles played against one another in a basketball game. At the end of the game, the 7 players on the Knights team shook hands with each of the 8 players on the Eagles team. How many handshakes were exchanged?

_____

22. Felix set a goal to increase his running speed by 1 minute per mile every 5 weeks. He starts out running 1 mile in 11 minutes. If he can accomplish his goal, how many weeks will it take him to run a mile in 8 minutes?

_____

23. A school store buys pencils in bulk and sells them to students for a small profit. The school store buys 100 pencils for $2.00. The store sells the pencils for 10¢ each. How many pencils would the store need to sell to make a total profit of $100?

_____

CHAPTER TEST

1. John has been working at Palazone's Pizza for 6 months. He works 2 afternoons per week and all day Saturday and Sunday for a total of 24 hours per week. He has been promised a 6% raise. How much will he make per week now?

   A. $ 48.00
   B. $144.00
   C. $288.00
   D. Not enough information is given.

2. Emily had $100.00 to spend. She spent $37.85 on clothes. How much change should she receive from $40.00?

   A. $ 2.15
   B. $ 3.15
   C. $60.00
   D. $62.15

3. Phil needed 52 eggs for the Spring Festival egg toss contest. He purchased the eggs at a store that only sold eggs by the dozen. How many dozen eggs did he purchase?

   A. 4
   B. 5
   C. 10
   D. Not enough information is given

4. Five beach volleyball teams play each other twice in a tournament. How many total volleyball matches are played?

   A. 10
   B. 20
   C. 40
   D. 50

5. Yvon made $6.50 per hour and worked 20 hours last week. How much gross pay did she make last week?

   A. $ 65.00
   B. $120.00
   C. $130.00
   D. Not enough information is given.

6. Elaine sells bread she bakes at home in three local food stores. She figured it is costing her 36¢ per loaf to make the bread, and she sells it for $1.25 per loaf. She works 25 hours a week.

   To find out how much Elaine makes an hour, you also need to know:

   A. how many days she works in a week.
   B. how many days a loaf of bread can stay on the shelf before the expiration date.
   C. how many loaves she sells in a week.
   D. how far away the stores are from her home.

7. The Mims family has a food budget of $500.00 per month. The first week of the month, they spent $135.32, the second week, $98.67, and the third week, $157.25. If there are 4 weeks in the month, how much money do they have budgeted for the last week?

   A. $391.24
   B. $108.76
   C. $266.01
   D. Not enough information is given.

8. Cindy went to look at puppies at the pet store. The store had the following puppies: a Cocker Spaniel, a German shepherd, a St. Bernard, and a Mastiff. Cindy found out that a Mastiff grows larger than a St. Bernard, and a German shepherd grows larger than a Cocker Spaniel but not as large as a St. Bernard. If Cindy wants to buy the puppy that will grow the largest, which breed should she pick?

   A. Cocker Spaniel
   B. German shepherd
   C. St. Bernard
   D. Mastiff

9. A hotel elevator has a weight capacity of 1000 pounds. The following people are waiting to get on the elevator.

    Joey weighs 232 pounds
    Felicia weighs 125 pounds
    Hector weighs 312 pounds
    Oscar weighs 155 pounds
    Terry weighs 251 pounds

    Which people should get on the elevator to get as close to the weight limit without going over?

    A. Joey, Felicia, Hector, and Oscar
    B. Joey, Hector, Oscar, and Terry
    C. Joey, Felicia, Oscar, and Terry
    D. Joey, Felicia, Hector, and Terry

10. Lyle spends the morning surfing in the ocean. He paddles out on his surfboard 50 yards and then catches a breaking wave to surf back to shore. Lyle can paddle out from shore for 10 yards before a wave pushes him back 2 yards. How many total yards does Lyle have to paddle to reach 100 yards from the shore?

    A. 40
    B. 50
    C. 60
    D. 70

11. At the farmer's market, a tomato farmer, a corn farmer, a green bean farmer, a squash farmer, and a peach farmer decided to trade among themselves. Each farmer exchanged a bushel of his product with the other farmers. How many bushels of tomatoes did the tomato farmer give out?

    A. 4
    B. 5
    C. 10
    D. 20

12. Five friends form a club and develop their own secret handshake. The handshake involves 2 hi-fives. How many hi-fives are exchanged if each of the 5 club members gives every other member the secret handshake?

    A. 5
    B. 10
    C. 20
    D. 40

13. At Wills High School, 40% of the students ride the bus to school. There are 660 students enrolled at Wills High School and 180 of them are freshmen. There are 150 seniors graduating this year. How many students ride the bus to school?

    A. 60
    B. 72
    C. 204
    D. 264

14. Lee works part time at a fast-food restaurant and makes $45 per week. He saves 75% towards college. How much will he have saved after 32 weeks?

    A. $1080.00
    B. $ 33.75
    C. $ 53.33
    D. Not enough information is given.

15. Renada had 2 twenty dollar bills, 4 ten dollar bills, 5 five dollar bills, and 11 one dollar bills. She used seven bills to pay a beautician $52 for a permanent. How many ten dollar bills did she use?

    A. 1
    B. 2
    C. 3
    D. 4

16. A 10 lb bag of fertilizer will cover 50 square feet of garden space. How many 10 lb bags of fertilizer would you need to cover a garden that is 480 square feet?

    A. 9
    B. 10
    C. 20
    D. 48

17. Five friends want to know who is the strongest arm wrestler. Each friend arm wrestles the others one time. How many times does each friend arm wrestle?

    A. 1
    B. 4
    C. 5
    D. 10

18. A player starts at Level 1 of a video game. He progresses to Level 2, Level 3, and Level 4. After completing Level 4, he gets sent back to Level 3. He then has to re-play Levels 3 and 4 before he can progress to Level 5. After playing Levels 3, 4, 5, and 6, he gets sent back to Level 5. This pattern continues throughout the game. How many total levels will he play before he gets to Level 12 for the first time?

    A. 19
    B. 20
    C. 21
    D. 22

19. Darlene has 10 coins that total $1.57 in her change holder. She has a mix of quarters, nickels, dimes, and pennies. How many quarters must she have in her change holder?

    A. 4
    B. 5
    C. 6
    D. 7

20. A student from America, a student from Mexico, a student from France, and a student from Italy meet at an international fair. Each student exchanges two kinds of coins from his own country with each of the other students. How many total coins are exchanged?

    A. 6
    B. 12
    C. 24
    D. 48

21. Carl, Kayla, Sean, Pete, Amanda, and Billy form a circle. Sean is beside Amanda, and Amanda is beside Kayla. Pete is between Carl and Kayla. Who is Billy between?

    A. Kayla and Amanda
    B. Pete and Carl
    C. Sean and Amanda
    D. Carl and Sean

22. Seven basketball players on the Timberwolves team shook hands with each of the eight players on the Falcons team. How many total handshakes were exchanged?

    A. 7
    B. 8
    C. 42
    D. 56

23. Patrick, Russ, Lenny, Keith, and Dave played each other twice in chess. How many chess games did Dave play?

    A. 4
    B. 5
    C. 8
    D. 10

Chapter 8

USING EXPONENTS AND THE METRIC SYSTEM

UNDERSTANDING EXPONENTS

Sometimes it is necessary to multiply a number by itself one or more times. For example, a math problem may need to multiply 3×3 or $5 \times 5 \times 5 \times 5$. In these situations, mathematicians have come up with a shorter way of writing out this kind of multiplication. Instead of writing 3×3, you can write 3^2, or, instead of $5 \times 5 \times 5 \times 5$, 5^4 means the same thing. The first number is the **base**. The small, raised number is called the **exponent**. The exponent tells how many times the base should be multiplied by itself.

EXAMPLE 1: 6^3 ← exponent, ← base   This means multiply 6 three times: $6 \times 6 \times 6$

You also need to know two special properties of exponents:

1. Any base number raised to the exponent of 1 equals the base number.
2. Any base number raised to the exponent of 0 equals 1.

EXAMPLE 2: $4^1 = 4$ $10^1 = 10$ $25^1 = 25$
$4^0 = 1$ $10^0 = 1$ $25^0 = 1$

Rewrite the following problems using <u>exponents</u>.

Example: $2 \times 2 \times 2 = 2^3$

1. $4 \times 4 \times 4 =$ _____
2. $5 \times 5 \times 5 \times 5 =$ _____
3. $11 \times 11 =$ _____
4. $8 \times 8 =$ _____
5. $17 \times 17 \times 17 =$ _____
6. $12 \times 12 =$ _____
7. $1 \times 1 \times 1 \times 1 \times 1 =$ _____
8. $7 \times 7 \times 7 \times 7 \times 7 =$ _____
9. $100 \times 100 \times 100 =$ _____

Use your calculator to figure what number each number with an exponent represents.

Example: $2^3 = 2 \times 2 \times 2 = 8$

10. $2^5 =$ _____
11. $10^2 =$ _____
12. $8^3 =$ _____
13. $3^4 =$ _____
14. $25^1 =$ _____
15. $10^5 =$ _____
16. $15^0 =$ _____
17. $9^2 =$ _____
18. $3^3 =$ _____

Express each of the following numbers as a number with an exponent.

Example: $4 = 2 \times 2 = 2^2$

19. $32 =$ _____
20. $64 =$ _____ or _____
21. $1000 =$ _____
22. $27 =$ _____
23. $81 =$ _____ or _____
24. $121 =$ _____
25. $16 =$ _____ or _____
26. $8 =$ _____
27. $49 =$ _____

SQUARE ROOT

Just as working with exponents is related to multiplication, so finding square roots is related to division. In fact, the sign for finding the square root of a number looks similar to a division sign. The best way to learn about square roots is to look at examples.

EXAMPLES: This is a square root problem: $\sqrt{64}$
It is asking, "What is the square root of 64?"
It means, "What number multiplied by itself equals 64?"
The answer is 8. $8 \times 8 = 64$.

Find the square root of the following numbers

$\sqrt{36}$ $6 \times 6 = 36$ so $\sqrt{36} = 6$ $\sqrt{144}$ $12 \times 12 = 144$ so $\sqrt{144} = 12$

Find the square roots of the following numbers.

1. $\sqrt{49}$ _____
2. $\sqrt{81}$ _____
3. $\sqrt{25}$ _____
4. $\sqrt{16}$ _____
5. $\sqrt{121}$ _____
6. $\sqrt{625}$ _____
7. $\sqrt{100}$ _____
8. $\sqrt{289}$ _____
9. $\sqrt{196}$ _____
10. $\sqrt{36}$ _____
11. $\sqrt{4}$ _____
12. $\sqrt{900}$ _____
13. $\sqrt{64}$ _____
14. $\sqrt{9}$ _____
15. $\sqrt{144}$ _____

MIXED PRACTICE

Write with exponents.	Find the square root.	Write as whole numbers.
1. $2 \times 2 \times 2 \times 2 \times 2$ _____	6. $\sqrt{1}$ _____	11. 16 _____
2. $5 \times 5 \times 5$ _____	7. $\sqrt{169}$ _____	12. 2^1 _____
3. 3×3 _____	8. $\sqrt{49}$ _____	13. 5^2 _____
4. $6 \times 6 \times 6 \times 6$ _____	9. $\sqrt{256}$ _____	14. 10^2 _____
5. 8×8 _____	10. $\sqrt{400}$ _____	15. 4^3 _____

MULTIPLYING AND DIVIDING BY MULTIPLES OF TEN

Multiplying and dividing decimal numbers by multiples of ten is easy. To **multiply**, simply move the decimal point to the **right**, and to **divide**, move the decimal point to the **left**.

> **Rule for multiplying by multiples of ten:** Move the decimal point to the right the same number of spaces as zeros.

EXAMPLE 1: 5.43 × 1000

LONG METHOD

```
    5.43
  ×1000
    3000
   4000
   5000
  5430.00 = 5,430
```

SHORT METHOD

1. Copy the number: 5.43
2. Count the number of zeros in 1000. (3)
3. Move the decimal point to the right 3 places. You will need to add a zero.

5.430.   5.43 × 1000 = 5,430

> **Rule for dividing by multiples of ten:** Move the decimal point to the left the same number of space as zeros.

EXAMPLE 2: 72.1 ÷ 100

LONG METHOD

```
        0.0721
   100)7.21
        7 00
          210
          200
           100
           100
             0
```

SHORT METHOD

1. Copy the number: 7.21
2. Count the number of zeros in 100. (2)
3. Move the decimal point to the left 2 places. You will need to add a zero.

.07.21   7.21 ÷ 100 = 0.0721

Simplify the following problems.

1. 7.54 × 100 = _____
2. 24.95 ÷ 10 = _____
3. 0.627 × 1000 = _____
4. 935.4 ÷ 10 = _____
5. 13.4 × 1000 = _____
6. 1.975 × 100 = _____
7. 18.6 ÷ 10,000 = _____
8. 0.59 ÷ 10 = _____
9. 17.25 ÷ 100 = _____
10. 54.9 × 1000 = _____
11. 0.3 × 100 = _____
12. 7.314 ÷ 10 = _____
13. 185.1 ÷ 100 = _____
14. 1.8 × 10,000 = _____
15. 26.5 ÷ 100 = _____
16. 0.625 ÷ 10 = _____
17. 8.05 × 100 = _____
18. 2.9 × 1000 = _____
19. 14.32 ÷ 10 = _____
20. 9.12 × 10 = _____
21. 400.4 ÷ 100 = _____
22. 0.418 ÷ 10 = _____
23. 3.952 × 100 = _____
24. 12.68 × 10 = _____

THE METRIC SYSTEM

The metric system uses units based on multiples of ten. The basic units of measure in the metric system are the **meter**, the **liter**, and the **gram**. Metric prefixes tell what multiple of ten the basic unit is multiplied by. Below is a chart of metric prefixes and their values. The ones rarely used are shaded.

Prefix	kilo (k)	hecto (h)	deka (da)	unit (m, L, g)	deci (d)	centi (c)	milli (m)
Meaning	1000	100	10	1	0.1	0.01	0.001

Multiply when changing from a greater unit to a smaller one; **divide** when changing from a smaller unit to a larger one. **The chart is set up to help you know how far and which direction to move a decimal point when making conversions from one unit to another.**

UNDERSTANDING METERS

The basic unit of **length** in the metric system is the **meter**. Meter is abbreviated "m".

Metric Unit	Abbreviation	Memory Tip	Equivalents
1 millimeter	mm	Thickness of a dime	10 mm = 1 cm
1 centimeter	cm	Width of the tip of the little finger	100 cm = 1 m
1 meter	m	Distance from the nose to the tip of fingers (a little longer than a yard)	1000 m = 1 km
1 kilometer	km	A little more than half a mile	

UNDERSTANDING LITERS

The basic unit of **liquid volume** in the metric system is the **liter**. Liter is abbreviated "L".

The liter is the volume of a cube measuring 10 cm on each side. A milliliter is the volume of a cube measuring 1 cm on each side. A capital L is used to signify liter, so it is not confused with the number 1.

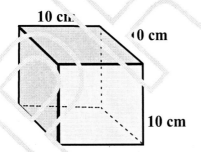

Volume = 1000 cm³ = 1 Liter
(a little more than a quart)

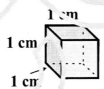

Volume = 1 cm³ = 1 mL
(an eyedropper holds 1 mL)

UNDERSTANDING GRAMS

The basic unit of **mass** in the metric system is the **gram**. Gram is abbreviated "g".

A **gram** is the **weight** of **one cubic centimeter** of **water** at $4°$ C.

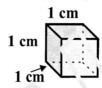

A large paper clip weighs about 1 gram (1g).
A nickel weighs 5 grams (5 g).
1000 grams = 1 kilogram (kg) = a little over 2 pounds

1 milligram (mg) = 0.001 gram. This is an extremely small amount and is used in medicine.

An aspirin tablet weighs 300 mg.

CONVERTING UNITS WITHIN THE METRIC SYSTEM

Converting units such as kilograms to grams or centimeters to decimeters is easy now that you know how to multiply and divide by multiples of ten.

Prefix	kilo (k)	hecto (h)	deka (da)	unit (m, L, g)	deci (d)	centi (c)	milli (m)
Meaning	1000	100	10	1	0.1	0.01	0.001

EXAMPLE 1: 2 L = _____ mL

2.000 L = 2000 mL

Look at the chart above. To move from liters to milliliters, you move to the right three places. So, to convert the 2 L to mL, move the decimal point three places to the right. You will need to add zeros.

EXAMPLE 2: 5.25 cm = _____ m

005.25 cm = 0.0525 m

To move from centimeters to meters, you need to move two spaces to the left. So, to convert 5.25 cm to m, move the decimal point two spaces to the left. Again, you need to add zeros.

Solve the following problems.

1. 35 mg = _____ g
2. 6 km = _____ m
3. 21.5 mL = _____ L
4. 4.9 mm = _____ cm
5. 5.35 kL = _____ mL
6. 32.1 mg = _____ kg
7. 156.4 m = _____ km
8. 25 mg = _____ cg
9. 17.5 L = _____ mL
10. 4.2 g = _____ kg
11. 0.06 daL = _____ dL
12. 0.417 kg = _____ cg
13. 18.2 cL = _____ L
14. 81.2 dm = _____ cm
15. 72.3 cm = _____ m
16. 0.003 kL = _____ L
17. 5.06 g = _____ mg
18. 1.058 mL = _____ cL
19. 43 hm = _____ km
20. 2.057 m = _____ cm
21. 564.3 g = _____ kg

ESTIMATING METRIC MEASUREMENTS

Choose the best estimates.

1. The height of an average man
   A. 18 cm
   B. 1.8 m
   C. 6 km
   D. 36 mm

2. The volume of a coffee cup
   A. 300 mL
   B. 20 L
   C. 5 L
   D. 1 kL

3. The width of this book
   A. 215 mm
   B. 75 cm
   C. 2 m
   D. 1.5 km

4. The weight of an average man
   A. 5 mg
   B. 15 cg
   C. 25 g
   D. 90 kg

5. The length of a basketball player's foot
   A. 2 m
   B. 1 km
   C. 30 cm
   D. 100 mm

6. The weight of a dime
   A. 3 g
   B. 30 g
   C. 10 cg
   D. 1 kg

7. The width of your hand
   A. 2 km
   B. 0.5 m
   C. 20 cm
   D. 90 mm

8. The length of a basketball court
   A. 1000 mm
   B. 250 cm
   C. 20 m
   D. 2 km

Choose the best units of measure.

9. The distance from Atlanta to Charlotte
   A. millimeter
   B. centimeter
   C. meter
   D. kilometer

10. The length of a house key
    A. millimeter
    B. centimeter
    C. meter
    D. kilometer

11. The thickness of a nickel
    A. millimeter
    B. centimeter
    C. meter
    D. kilometer

12. The width of a classroom
    A. millimeter
    B. centimeter
    C. meter
    D. kilometer

13. The length of a piece of chalk
    A. millimeter
    B. centimeter
    C. meter
    D. kilometer

14. The height of a peach tree
    A. millimeter
    B. centimeter
    C. meter
    D. kilometer

SCIENTIFIC NOTATION

Mathematicians use **scientific notation** to express very large and very small numbers. **Scientific notation** expresses a number in the following form:

only one digit before the decimal → $x.xx \times 10^x$ ← multiplied by a multiple of ten

remaining digits not ending in zeros after the decimal →

USING SCIENTIFIC NOTATION FOR LARGE NUMBERS

Scientific notation simplifies very large numbers that have many zeros. For example, Pluto averages a distance of 5,900,000,000 kilometers from the sun. In scientific notation, a decimal is inserted after the first digit (5.), the rest of the digits are copied except for the zeros at the end (5.9), and the result is multiplied by 10^9. The exponent = the total number of digits in the original number minus 1 or the number of spaces the decimal point moved.

$5,900,000,000 = 5.9 \times 10^9$ The following are more examples:

EXAMPLES: $32,560,000,000 = 3.256 \times 10^{10}$ $5,060,000 = 5.06 \times 10^6$

decimal moves 10 spaces to the left ↲   ↳ decimal moves 6 spaces to the left

Convert the following numbers to scientific notation.

1. 4,230,000,000 = _____
2. 64,300,000 = _____
3. 951,000,000,000 = _____
4. 12,200 = _____
5. 20,350,000,000 = _____
6. 9,000 = _____
7. 450,000,000,000 = _____
8. 6,200 = _____
9. 87,000,000 = _____
10. 105,000,000 = _____
11. 1,083,000,000,000 = _____
12. 204,000 = _____

To convert a number written in scientific notation back to conventional form, reverse the steps.

EXAMPLE: $4.02 \times 10^5 = 4.02000 = 402,000$ Move the decimal 5 spaces to the right and add zeros.

Convert the following numbers from scientific notation to conventional numbers.

13. $6.85 \times 10^8 =$ _____
14. $1.3 \times 10^{10} =$ _____
15. $4.908 \times 10^4 =$ _____
16. $7.102 \times 10^6 =$ _____
17. $2.5 \times 10^3 =$ _____
18. $9.114 \times 10^5 =$ _____
19. $5.87 \times 10^7 =$ _____
20. $8.047 \times 10^8 =$ _____
21. $3.81 \times 10^5 =$ _____
22. $9.5 \times 10^{12} =$ _____
23. $1.504 \times 10^6 =$ _____
24. $7.3 \times 10^9 =$ _____

USING SCIENTIFIC NOTATION FOR SMALL NUMBERS

Scientific notation also simplifies very small numbers that have many zeros. For example, the diameter of a helium atom is 0.000000000244 meters. It can be written in scientific notation as 2.44×10^{-10}. The first number is always greater than 0, and the first number is always followed by a decimal point. The negative exponent indicates how many digits the decimal point moved to the right. The exponent is negative when the original number is less than 1. To convert small numbers to scientific notation, follow the **EXAMPLES** below.

EXAMPLES: $0.00058 = 5.8 \times 10^{-4}$ $0.00003059 = 3.059 \times 10^{-5}$

decimal point moves 4 spaces to the right — negative exponent indicates the original number is less than 1.   decimal moves 5 spaces to the right

---

Convert the following numbers to scientific notation.

1. 0.00000254 = _____
2. 0.00000000508 = _____
3. 0.000008004 = _____
4. 0.00047 = _____
5. 0.000000005478 = _____
6. 0.00000059 = _____
7. 0.000000712 = _____
8. 0.00025 = _____
9. 0.000000501 = _____
10. 0.000006 = _____
11. 0.0000000000875 = _____
12. 0.00004 = _____

Now convert small numbers written in scientific notation back to conventional form.

EXAMPLE: $3.08 \times 10^{-5} = 00003.08 = 0.0000308$ Move the decimal 5 spaces to the left and add zeros.

Convert the following numbers from scientific notation to conventional numbers.

13. 1.18×10^{-7} = _____
14. 2.3×10^{-5} = _____
15. 6.705×10^{-9} = _____
16. 4.1×10^{-6} = _____
17. 7.032×10^{-4} = _____
18. 5.48×10^{-10} = _____
19. 2.75×10^{-8} = _____
20. 4.07×10^{-7} = _____
21. 5.2×10^{-3} = _____
22. 7.01×10^{-6} = _____
23. 4.4×10^{-5} = _____
24. 3.43×10^{-2} = _____

CHAPTER REVIEW

Rewrite the following problems using exponents.

1. $3 \times 3 \times 3 \times 3$ _____
2. $5 \times 5 \times 5$ _____
3. $10 \times 10 \times 10 \times 10 \times 10$ _____
4. 25×25 _____

Use a calculator to figure the solution to the following.

5. 2^2 = ____
6. 5^3 = ____
7. 12^1 = ____
8. 15^0 = ____
9. 10^4 = ____
10. 7^2 = ____

Find the square root.

11. $\sqrt{25}$ = ____
12. $\sqrt{16}$ = ____
13. $\sqrt{36}$ = ____
14. $\sqrt{100}$ = ____
15. $\sqrt{81}$ = ____
16. $\sqrt{49}$ = ____

Simplify the following problems.

17. 5.623×10 = _____
18. $245.6 \div 100$ = _____
19. 14.95×1000 = _____
20. $0.365 \times 10,000$ = _____
21. $0.0587 \div 1000$ = _____
22. $1.2 \div 100$ = _____

Solve the following problems.

23. 120 m = _____ km
24. 9 g = _____ mg
25. 0.02 kL = _____ L
26. 1.5 mg = _____ g
27. 15 cm = _____ mm
28. 5 L = _____ mL
29. 0.005 kg = _____ g
30. 55 mL = _____ L

Convert the following numbers to scientific notation.

31. 22,300,000 = _____
32. 5,340,000 = _____
33. 0.000000005874 = _____
34. 1,451 = _____
35. 0.000041 = _____
36. 0.004178 = _____
37. 105,000 = _____
38. 705,000,000 = _____
39. 0.000074 = _____
40. 0.08 = _____
41. 105 = _____
42. 0.0048754 = _____
43. 62,400 = _____

Convert the following numbers from scientific notation to conventional numbers.

44. 5.204×10^{-5} = _____
45. 1.02×10^{7} = _____
46. 8.1×10^{5} = _____
47. 2.0078×10^{4} = _____
48. 4.7×10^{-3} = _____
49. 7.75×10^{-8} = _____
50. 9.785×10^{9} = _____
51. 3.51×10^{2} = _____
52. 6.32514×10^{3} = _____
53. 1.584×10^{-6} = _____
54. 7.041×10^{4} = _____
55. 4.09×10^{-7} = _____
56. 3×10^{-10} = _____

CHAPTER TEST

1. $5^3 =$

    A. 15
    B. 25
    C. 125
    D. 555

2. $\sqrt{64} =$

    A. −64
    B. 4
    C. 8
    D. 32

3. $2^4 =$

    A. 2
    B. 4
    C. 8
    D. 16

4. $\sqrt{100} =$

    A. 1
    B. 10
    C. 100
    D. 10,000

5. $2354^0 =$

    A. 0
    B. 1
    C. 2354
    D. 23540

6. $4.0587 \times 1000 =$

    A. 40.587
    B. 405.87
    C. 4058.7
    D. 40587

7. $65.089 \div 10,000 =$

    A. 6.5089
    B. 0.65089
    C. 0.065089
    D. 0.0065089

8. $0.0084 \times 100 =$

    A. 0.084
    B. 0.84
    C. 84
    D. 840

9. $0.009 \div 10 =$

    A. 0.0009
    B. 0.09
    C. 0.9
    D. 9.0

10. $0.00058 \times 100,000 =$

    A. 58
    B. 0.0000000058
    C. 058
    D. 580,000

11. Which of the following choices is the best estimate for the width of a standard door?

    A. 9 meters
    B. 9 kilometers
    C. 90 centimeters
    D. 90 millimeters

12. Which of the following choices is the best estimate for the weight of a house cat?

    A. 4 kilograms
    B. 40 kilograms
    C. 40 grams
    D. 400 milligrams

13. Which of the following choices is the best estimate for the volume of a carton of orange juice?

    A. 2 liters
    B. 2 kiloliters
    C. 20 milliliters
    D. 20 liters

14. Which unit of measure would be most appropriate to measure the dimensions of a classroom?

    A. millimeters
    B. centimeters
    C. meters
    D. kilometers

15. Which of the following is equal to 30 millimeters?

    A. 0.03 meters
    B. 0.003 kilometers
    C. 300 centimeters
    D. 30,000 meters

16. Which of the following is equal to 5 L?

    A. 0.005 mL
    B. 0.5 kL
    C. 5,000 mL
    D. 5,000 kL

17. Which of the following is equal to 300 kg?

    A. 0.3 g
    B. 0.0003 mg
    C. 300,000 g
    D. 30,000,000 mg

18. Which of the following is equal to 7.054×10^5?

    A. 0.0000705
    B. 0.00007054
    C. 70,540
    D. 705,400

19. Which of the following is equal to 1.02×10^{-4}?

    A. 102
    B. 1,020
    C. 0.000102
    D. 0.0000102

20. Which of the following is equal to 6.5×10^{10}?

    A. 0.000000000065
    B. 0.00000000065
    C. 65,000,000,000
    D. 650,000,000,000

21. Which of the following is equal to 9.01×10^{-7}?

    A. 90,100,000
    B. 9,010,000,000
    C. 0.000000901
    D. 0.0000000901

22. What is 7,003,780 expressed in scientific notation?

    A. 7.00378×10^5
    B. 7.00378×10^{-5}
    C. 7.00378×10^6
    D. 7.00378×10^{-6}

23. What is 0.0000000042 expressed in scientific notation?

    A. 5.42×10^8
    B. 5.42×10^{-8}
    C. 5.42×10^9
    D. 5.42×10^{-9}

24. How should you express 40009.2 in scientific notation?

    A. 4.00092×10^4
    B. 4.00092×10^{-4}
    C. 4.00092×10^6
    D. 4.00092×10^{-6}

25. How should you express 0.000001 in scientific notation?

    A. 1×10^5
    B. 1×10^{-5}
    C. 1×10^6
    D. 1×10^{-6}

Chapter 9

CUSTOMARY MEASUREMENTS

USING THE RULER

Practice measuring the objects below with a ruler.

Measure these distances in inches.

1. How tall is the calculator? _____
2. How wide is the maple leaf? _____
3. How long is the car? _____
4. How far is it from the nose of the airplane to the nose of the camel? _____
5. How long is the trumpet? _____
6. How far is it from the middle of the bicycle's back wheel to the middle of the front wheel? _____
7. How long is the hour hand on the clock? _____
8. How far is it from the nose of the car to the mouthpiece of the trumpet? _____

Measure these distances in centimeters.

9. How long is the minute hand of the clock? _____
10. How tall is the maple leaf? _____
11. How tall is the camel? _____
12. How wide is the calculator? _____
13. How long is the plane? _____
14. How far is it from the tip of the hour hand to the tallest hump on the camel's back? _____
15. How far is it from the tip of the camel's nose to its front hoof? _____
16. How wide is the airplane from wing tip to wingtip? _____

MORE MEASURING

Measure the following line segments.

1. ———————————— = _____ in   6. ———————————— = _____ cm

2. ——————— = _____ in   7. ——————— = _____ cm

3. —————————— = _____ in   8. ——————— ——— = _____ cm

4. ———————————— = _____ in   9. ——————— — — = _____ cm

5. ——————— = _____ in   10. ——— ——— = _____ cm

Measure the dimensions of the following figures.

11.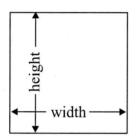

    height: _____ cm

    width: _____ cm

12.

    height from the top of the stem to the bottom.
    _____ in

13.

    width of the desk top:
    _____ in

    height of the desk:
    _____ in

14.

    length of guitar:
    _____ cm

15.

    width of soccer ball:
    _____ cm

16.

    height of tent: _____ in

    width of tent at its base:
    _____ in

CONVERTING UNITS OF MEASURE

English System of Measure:

Time	Abbreviations
1 week = 7 days	week = wk
1 day = 24 hours	hour = hr or h
1 hour = 60 minutes	minutes = min
1 minute = 60 seconds	seconds = sec

Length	Abbreviations
1 mile = 5,280 feet	mile = mi
1 yard = 3 feet	yard = yd
1 foot = 12 inches	foot = ft

Volume	Abbreviations
1 gallon = 4 quarts	gallon = gal
1 quart = 2 pints	quart = qt
1 pint = 2 cups	pint = pt
1 cup = 8 ounces	ounce = oz

Weight	Abbreviations
16 ounces = 1 pound	pound = lb
	ounce = oz

EXAMPLE: Simplify: 2 days 34 hr 75 min

Step 1 75 minutes is more than 1 hour. There are 60 minutes in an hour so divide 75 by 60.

$$60\overline{)75} \rightarrow 1\text{ hr remainder }15\text{ min}$$

```
         2 days  34 hr  75 min
       +          1 hr  15 min
         2 days  35 hr  15 min
```

Step 2 35 hours is more than 1 day. There are 24 hours in a day so divide 35 hours by 24.

$$24\overline{)35} \rightarrow 1\text{ day remainder }11\text{ hr}$$

```
         2 days  35 hr  15 min
       + 1 day   11 hr
         3 days  11 hr  15 min
```

Simplify the following:

1. 3 lb 20 oz

2. 2 cup 12 oz

3. 3 wk 9 days 30 hr

4. 1 pt 1 cup 16 oz

5. 2 hr 84 min 62 sec

6. 1 gal 6 qt 3 pt

7. 3 yd 10 ft 18 in

8. 2 wk 8 days 36 hours

9. 2 ft 18 in

10. 1 lb 33 oz

11. 23 hr 62 min 94 sec

12. 3 days 54 hr 75 min

ADDING UNITS OF MEASURE

EXAMPLE: Add 3 days 8 hours 45 minutes + 2 days 20 hours 35 minutes.

Step 1 Arrange the numbers so that like units are in the same column.

Step 2 Add

```
  3 days   8 hr  45 min
+ 2 days  20 hr  35 min
  5 days  28 hr  80 min
```

Step 3 Simplify the answer.

80 minutes =
```
  5 days  28 hr  80 min
                1 hr  20 min
  5 days  29 hr  20 min
```

29 hours =
```
          1 day   5 hr
  6 days   5 hr  20 min
```

Add and simplify the answers.

1. 2 hr 50 min + 15 hr 15 min

4. 4 days 22 hr + 7 days 8 hr

2. 4 lb 14 oz + 6 lb 13 oz

5. 5 gal 2 qt + 1 gal 3 qt

3. 2 ft 8 in + 2 ft 11 in

6. 3 yd 2 ft + 1 yd 2 ft

7. 3 gal 2 qt 1 pt
  + 1 gal 3 qt 1 pt

10. 2 hr 48 min 30 sec
  + 1 hr 20 min 45 sec

13. 1 pt 1 cup 6 oz
  + 1 pt 1 cup 4 oz

8. 3 mi 3,745 ft
  + 2 mi 3,472 ft

11. 1 yd 2 ft 7 in
  + 4 yd 1 ft 8 in

14. 5 days 13 hr 20 min
  + 1 day 15 hr 45 min

9. 5 wk 4 days 20 hr
  + 2 wk 5 days 15 hr

12. 3 days 5 hr 12 min
  + 5 days 20 hr 50 min

15. 2 gal 3 qt 1 cup
  + 5 gal 2 qt 1 cup

SUBTRACTING UNITS OF MEASURE

EXAMPLE: 10 ft 5 in − 6 ft 7 in

Step 1 Arrange the numbers so that like units are in the same column.

```
  10 ft  5 in
−  6 ft  7 in
```

Step 2 Subtract like units by borrowing when necessary.

```
   9   17
  10 ft  5 in
−  6 ft  7 in
   3 ft 10 in
```

You must borrow 1 foot to subtract. 1 foot = 12 inches
You need to add the 12 inches to the 5 inches you already have (17) and then subtract.

Subtract by borrowing when necessary.

1. 2 days 3 hr − 1 day 20 hr

4. 5 gal 1 qt − 3 gal 2 qt

2. 4 hr 13 min − 1 hr 30 min

5. 3 cups 5 oz − 1 cup 6 oz

3. 3 wk 2 days − 2 wk 5 days

6. 13 min 34 sec − 6 min 50 sec

7. 6 yd 1 ft 5 in
    − 2 yd 3 ft 7 in

10. 8 ft 4 in
    − 3 ft 6 in

13. 1 yd 3 ft 7 in
    −   5 ft 10 in

8. 25 lb 11 oz
    − 13 lb 15 oz

11. 10 gal 2 qt
    − 2 gal 3 qt

14. 1 pt 1 cup 3 oz
    −   1 cup 5 oz

9. 8 hr 15 min 45 sec
    − 3 hr 23 min 40 sec

12. 6 days 14 hr 35 min
    − 3 days 7 hr 50 min

15. 15 yd 2 ft 8 in
    − 4 yd   9 in

MULTIPLYING UNITS OF MEASURE

EXAMPLE:
$$\begin{array}{r} 4 \text{ lb } 9 \text{ oz} \\ \times 3 \\ \hline \end{array}$$

Step 1 Multiply each column separately.
$$\begin{array}{r} 4 \text{ lb } 9 \text{ oz} \\ \times 3 \\ \hline 12 \text{ lb } 27 \text{ oz} \end{array}$$

Step 2 Simplify the answer.

27 oz is more than 1 lb. There are 16 oz in a pound. Divide 27 by 16.

$$\begin{array}{r} 1 \text{ lb} \\ 16\overline{)27} \\ -16 \\ \hline 11 \text{ oz} \end{array}$$

$$\begin{array}{r} 4 \text{ lb } 9 \text{ oz} \\ \times 3 \\ \hline 12 \text{ lb } 27 \text{ oz} \\ 1 \text{ lb } 11 \text{ oz} \\ \hline 13 \text{ lb } 11 \text{ oz} \end{array}$$

Multiply the following and simplify

1. 11 ft × 5
2. 3 ft 4 in × 5
3. 1 gal 3 qt × 5
4. 5 ft 5 in × 3
5. 5 hr 15 min × 4
6. 2 yd 2 ft × 8
7. 2 hr 22 min × 3
8. 4 min 40 sec × 5
9. 1 hr 20 min × 4
10. 5 lb 3 oz × 9
11. 1 cup 5 oz × 4
12. 9 lb 2 oz × 8

13. $\begin{array}{r} 3 \text{ lb } 6 \text{ oz} \\ \times \\ \hline \end{array}$

14. $\begin{array}{r} 5 \text{ yd } 8 \text{ ft } 2 \text{ in} \\ \times 6 \\ \hline \end{array}$

15. $\begin{array}{r} 9 \text{ hr } 15 \text{ min} \\ \times 4 \\ \hline \end{array}$

16. $\begin{array}{r} 1 \text{ cup } 3 \text{ oz} \\ \times 6 \\ \hline \end{array}$

17. $\begin{array}{r} 30 \text{ min } 20 \text{ sec} \\ \times 2 \\ \hline \end{array}$

18. $\begin{array}{r} 4 \text{ hr } 10 \text{ min} \\ \times 7 \\ \hline \end{array}$

19. $\begin{array}{r} 2 \text{ days } 5 \text{ hr} \\ \times 6 \\ \hline \end{array}$

20. $\begin{array}{r} 50 \text{ ft } 6 \text{ in} \\ \times 4 \\ \hline \end{array}$

21. $\begin{array}{r} 4 \text{ lb } 9 \text{ oz} \\ \times 5 \\ \hline \end{array}$

CHAPTER REVIEW

Measure the following distances:

1. The length of the pencil _____ in
2. The width of the bell _____ in
3. The width of the pencil _____ cm
4. The distance between the point of the pencil and the top of the sail. _____ cm

Measure the following line segments:

5. ──────────── _____ in
6. ──────────── _____ cm
7. ──── _____ in
8. ──── _____ cm

Measure the dimensions of the following figures:

9.

   length: _____ in

11.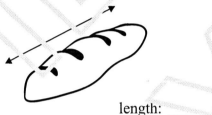

    length: _____ cm

10.

    height: _____ in

12.

    height: _____ cm

Simplify the following problems.

13.   2 days 3 hr 25 min
    + 3 days 22 hr 40 min

16.   5 yd 1 ft 7 in
    − 1 yd 2 ft 5 in

19.   3 lb 5 oz
    ×   6

14.   5 hr 15 min 20 sec
    ×   4

17.   5 gal 2 qt 1 cup
    + 1 gal 3 qt 1 cup

20.   5 cups 5 oz
    − 3 cups 7 oz

15.   5 gal 1 qt 1 pt
    − 2 gal 3 qt 1 pt

18.   2 days 15 hr 24 min
    ×   4

21.   4 lb 9 oz
    + 2 lb 11 oz

CHAPTER TEST

1. Using a ruler, measure the length of the line segment below in inches.

   _____

   A. $1\frac{3}{8}$ inches
   B. $1\frac{7}{8}$ inches
   C. $2\frac{3}{8}$ inches
   D. $2\frac{5}{8}$ inches

2. Using a ruler, what are the dimensions of the shell pictured below?

   A. $\frac{1}{2}$ inch wide and $\frac{1}{2}$ inch tall
   B. $\frac{1}{2}$ inch wide and 1 inch tall
   C. $\frac{3}{4}$ inch wide and $\frac{3}{4}$ inch tall
   D. 1 inch wide and 1 inch tall

3. With a ruler, determine the length of the push pin pictured below.

   A. $\frac{3}{4}$ inch
   B. 1 inch
   C. $1\frac{1}{4}$ inches
   D. $2\frac{1}{2}$ inches

4. Use a ruler to determine the height of the snowman below in inches.

   A. $\frac{3}{8}$ inch
   B. $\frac{3}{4}$ inch
   C. $\frac{1}{2}$ inch
   D. $\frac{7}{8}$ inch

5. With a ruler, measure the length of the line segment below.

   _____

   A. 3 cm
   B. 3.5 cm
   C. 3.8 cm
   D. 4.1 cm

6. With a ruler, determine the length of the following figure.

   A. 2 cm
   B. 3 cm
   C. 4 cm
   D. 5 cm

7. Use a ruler to determine the width of the bowling ball.

   A. 1.3 cm
   B. 3.2 cm
   C. 3.7 cm
   D. 4.9 cm

8. What is the length of the line below?

   _____

   A. 2.3 cm
   B. 4.5 cm
   C. 5.2 cm
   D. 5.7 cm

9. Simplify: 4 hours 62 minutes 93 seconds

   A. 4 hours 2 minutes 33 seconds
   B. 4 hours 3 minutes 33 seconds
   C. 5 hours 3 minutes 33 seconds
   D. 5 hours 4 minutes 33 seconds

10. 2 yd 2 ft 3 in + 3 yd 1 ft 10 in

    A. 5 yd 1 ft 1 in
    B. 5 yd 2 ft 1 in
    C. 6 yd 1 ft 1 in
    D. 6 yd 1 in

11. 4 lb 7 oz − 1 lb 3 oz

    A. 3 lb 4 oz
    B. 3 lb 10 oz
    C. 5 lb 10 oz
    D. 6 lb 2 oz

12. 2 gal 3 qt 1 pint × 3

    A. 6 gal 3 qt 1 pt
    B. 8 gal 2 qt 1 pt
    C. 10 gal 1 qt 1 pt
    D. 11 gal 1 pt

13. Simplify: 1 week 8 days 50 hours

    A. 2 weeks 2 days 2 hours
    B. 2 weeks 3 days 2 hours
    C. 2 weeks 5 days 2 hours
    D. 3 weeks 2 days

14. 3 lb 6 oz − 2 lb 10 oz

    A. 1 lb 4 oz
    B. 1 lb 12 oz
    C. 8 oz
    D. 12 oz

15.   1 day 2 hr 15 min
    −   1 hr 30 min

    A.   45 min
    B.   1 hr 45 min
    C. 1 day   45 min
    D. 1 day 1 hr 15 min

16.   8 yd 2 ft 9 in
    + 1 yd 2 ft 10 in

    A. 9 yd 1 ft 7 in
    B. 10 yd 1 ft 7 in
    C. 10 yd 2 ft 7 in
    D. 11 yd   7 in

17.   1 gal 1 pt
    ×   5

    A. 5 gal 1 qt 1 pt
    B. 5 gal 2 qt 1 pt
    C. 6 gal 1 qt 1 pt
    D. 6 gal   1 pt

18.   3 hr 25 min 34 sec
    + 1 hr 34 min 30 sec

    A. 4 hr 59 min 4 sec
    B. 5 hr   4 sec
    C. 5 hr 1 min 4 sec
    D. 6 hr   4 sec

19.   5 yd 1 ft 3 in
    − 2 yd 2 ft 6 in

    A. 2 yd 1 ft 9 in
    B. 2 yd 1 ft 6 in
    C. 3 yd 2 ft 3 in
    D. 8 yd   9 in

20.   4 minutes 22 seconds
    − 2 minutes 45 seconds

    A. 1 minute 37 seconds
    B. 1 minute 23 seconds
    C. 2 minutes 17 seconds
    D. 2 minutes 23 seconds

Chapter 10

DATA INTERPRETATION

READING TABLES

A **table** is a concise way to organize large quantities of information using rows and columns. **Read each table carefully and then answer the questions that follow.**

Some employers use a tax table like the one below to figure how much Federal Income Tax should be withheld from a single person paid weekly. The number of withholding allowances claimed is also commonly referred to as the number of deductions claimed.

Federal Income Tax Withholding Table					
SINGLE Persons – WEEKLY Payroll Period					
If the wages are –		And the number of withholding allowances claimed is –			
		0	1	2	3
At least	But less than	The amount of income tax to be withheld is –			
$250	260	21	23	16	9
$260	270	32	25	17	10
$270	280	34	26	19	12
$280	290	35	28	20	13
$290	300	37	29	22	15

1. David is single, claims 2 withholding allowances and earned $275 last week. How much Federal Income Tax was withheld? _____

2. Cecily earned $291 last week and claims 0 deductions. How much Federal Income Tax was withheld? _____

3. Sheri claims 3 deductions and earned $268 last week. How much Federal Income Tax was withheld from her check? _____

4. Mitch is single and claims 1 allowance. Last week he earned $291. How much was withheld from his check for Federal Income Tax? _____

5. Ginger earned $275 this week and claims 0 deductions. How much Federal Income Tax will be withheld from her check? _____

6. Bill is single and earns $263 per week. He claims 1 withholding allowance. How much Federal Income Tax is withheld each week? _____

Nutritional tables appear on the packaging of nearly every food we eat. Use the nutritional table below to answer questions 1-6.

Peanut Butter Spread				
Nutrition Facts	Amount/serving		%DV*	Vitamin A 0%
Serv. Size 2 tbsp (36 g)	**Total Fat**	12 g	18%	Vitamin C 0%
Servings Per Container about 14	Sat. Fat	2.5 g	12%	Calcium 2%
	Cholest.	0 mg	0%	Iron 4%
Calories 190	**Sodium**	210 mg	9%	Niacin 25%
Calories from Fat 100	**Total Carb.**	5 g	5%	Vitamin B_6 6%
* Percent Daily Values (DV) are based on a 2,000 calorie diet	Fiber	1 g	5%	Folate 6%
	Sugars	5 g		Magnesium 15%
	Protein	9 g		Zinc 0%
				Copper 10%

1. How many calories are in 1 tablespoon of peanut butter spread? _____

2. How many milligrams of sodium are in 4 tablespoons of peanut butter spread? _____

3. How many grams of saturated fat are in 6 tablespoons of peanut butter spread? _____

4. If you eat 1 tablespoon of peanut butter spread, what percent of daily value of copper would you eat? _____

5. If you ate 4 tablespoons of peanut butter spread, how many grams of protein would you get? _____

6. If you used 2 tablespoons of peanut butter spread to make a peanut butter and jelly sandwich, how many grams of sugar came from the peanut butter spread? _____

Use the following life expectancy table to answer questions 7-12.

Average Life Expectancy		
Country	Male	Female
Bolivia	45	51
United Kingdom	72	77
India	46	43
Japan	74	81
Peru	52	55
Sweden	73	80
USA	71	78

7. Where do males live longer than females? _____

8. Where do women live the longest? _____

9. How much longer do men in the USA live than the men in Peru? _____

10. Where do people live the longest? _____

11. How much longer does a female in Sweden live than a male in Bolivia? _____

12. How much shorter is the average life of a female in the USA than a female in Japan? _____

MILEAGE CHART

A mileage chart will tell you the distance between two cities. To read the chart, find the city you start from along the side of the chart. Find the city of your destination along the top. Then read down the column and across the row to see where they intersect. The box where they intersect will tell how many miles it is between the two cities.

	Atlanta	Boise	Columbia	Dallas	Denver	Jacksonville	Kansas City	Little Rock	Miami	New Orleans	Reno	Seattle
Atlanta	0	1750	200	750	1125	275	650	462	550	400	1875	2075
Boise	1750	0	1925	1200	625	2000	1075	1300	2200	1875	300	400
Columbia	200	1925	0	975	1300	250	800	650	550	600	2050	2200
Dallas	750	1200	975	0	625	950	500	350	1100	500	1250	1600
Denver	1125	625	1300	625	0	1400	500	700	1600	1025	750	1000
Jacksonville	275	2000	250	950	1400	0	950	700	300	500	2125	2350
Kansas City	650	1075	800	500	500	950	0	300	1200	700	1225	1400
Little Rock	462	1300	650	350	700	700	300	0	900	400	1425	1675
Miami	550	2200	550	1100	1600	300	1200	900	0	600	2350	2600
New Orleans	400	1875	600	500	1025	500	700	400	600	0	1725	2025
Reno	1875	300	2050	1250	750	2125	1225	1425	2350	1725	0	550
Seattle	2075	400	2200	1600	1000	2350	1400	1675	2600	2025	550	0

Find the distance between the following cities.

1. Little Rock to Dallas _____
2. Seattle to Jacksonville _____
3. Miami to Seattle _____
4. Kansas City to New Orleans _____
5. Denver to Dallas _____
6. Reno to Boise _____
7. Atlanta to Reno _____
8. Little Rock to Columbia _____
9. Seattle to Kansas City _____
10. Columbia to Reno _____
11. Jacksonville to Little Rock _____
12. Columbia to Denver _____
13. New Orleans to Miami _____
14. Reno to Jacksonville _____
15. Miami to Boise _____
16. Jacksonville to Denver _____
17. Kansas City to Columbia _____
18. Dallas to Seattle _____
19. Denver to Miami _____
20. Atlanta to Seattle _____

BAR GRAPHS

Bar graphs can be either vertical or horizontal. There may be just one bar or more than one bar for each interval. Sometimes each bar is divided into two or more parts. In this section, you will work with a variety of bar graphs. Be sure to read all titles, keys, and labels to completely understand all the data that is presented. **Answer the questions about each graph below.**

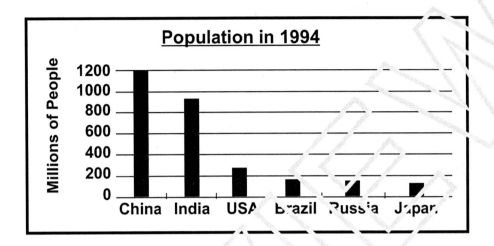

1. Which country has over 1 billion people? _____

2. How many countries have fewer than 200,000,000 people? _____

3. How many more people does India have than Japan? _____

4. If you added together the populations of the USA, Brazil, Russia and Japan, would it come closer to the population of India or China? _____

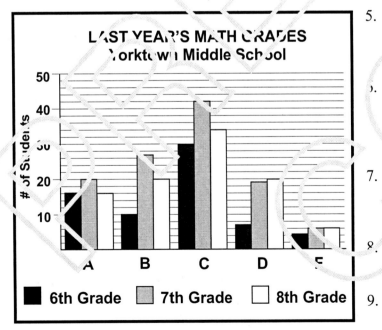

5. How many of last year's 6th graders made C's in math? _____

6. How many more math students made B's in the 7th grade than in the 8th grade? _____

7. How many students at Yorktown Middle school made D's in math last year? _____

8. How many 8th graders took math last year? _____

9. How many students made A's in math last year? _____

150 Copyright © American Book Company

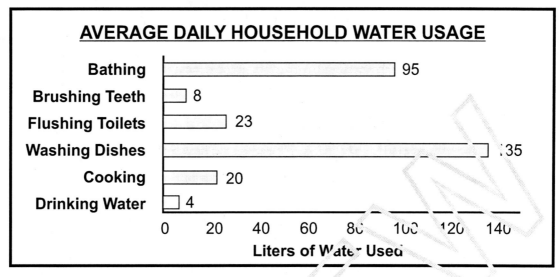

1. How many liters of water does the average household use every day by washing dishes and cooking? _____

2. How many liters of water would an average household drink in a week (7 days)? _____

3. How many liters of water does an average household use in a day? _____

4. How many more liters does an average household use for bathing than it does for cooking? _____

5. How many liters does the average household use every day to bathe and brush teeth? _____

An elementary school planted four trees: a dogwood, a maple, a pine, and an oak. All four trees were 36 inches tall when planted. The students monitored the growth of the trees each season for one year. Study the graph below, and then answer the questions.

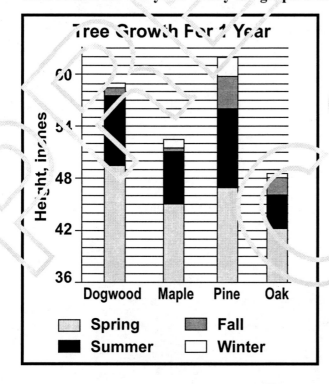

6. How many more inches did the pine tree grow than the oak tree in the summer? _____

7. Which tree grew the most during the winter? _____

8. Which tree grew the least during the fall? _____

9. How tall was the dogwood after one year? _____

10. How many more inches did the maple tree grow than the oak tree during the whole year? _____

LINE GRAPHS

Line graphs often show how data changes over time. Study the line graph below charting temperature changes for a day in Atlanta, Georgia. Then answer the questions that follow.

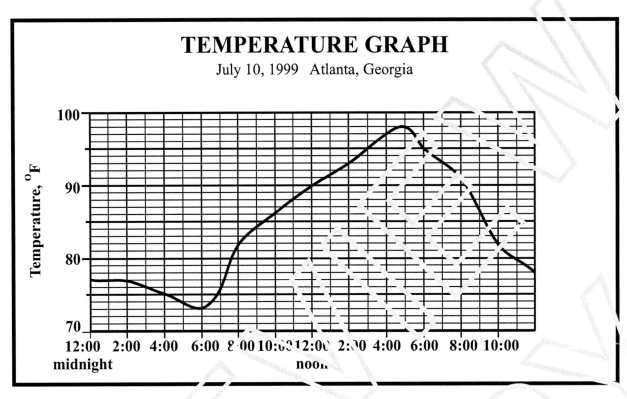

Study the graph and then answer the questions below.

1. When was the coolest time of the day? _____

2. When was the hottest time of the day? _____

3. How much did the temperature rise between 6:00 a.m and 2:00 p.m.? _____

4. How much did the temperature drop between 6:00 p.m. and 11:00 p.m.? _____

5. What is the difference in temperature between 8:00 a.m. and 8:00 p.m.? _____

6. Between which two hours was the greatest increase in temperature? _____

7. Between which hours of the day did the temperature continually increase? _____

8. Between which two hours of the day did the temperature change the least? _____

9. How much did the temperature decrease from 2:00 a.m. to 6:00 a.m.? _____

10. During which two times of day was the temperature 93°F? _____

MULTIPLE LINE GRAPHS

Multiple line graphs are a way to present a large quantity of data in a small space. It would often take several paragraphs to explain in words the same information that one graph could do.

On the graph below, there are three lines. You will need to read the **key** to understand the meaning of each.

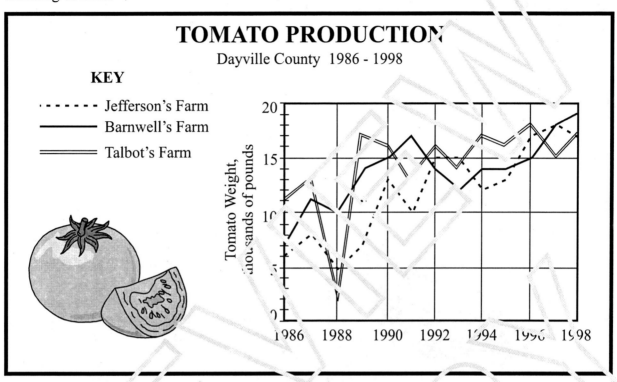

Study the graph, and then answer the questions below.

1. In what year did Barnwell's Farm produce 8,000 pounds of tomatoes more than Talbot's Farm? _____

2. In which year did Dayville County produce the most pounds of tomatoes? _____

3. How many more pounds of tomatoes did Barnwell's Farm produce compared to Talbot's Farm in 1991? _____

4. How many pounds of tomatoes did Dayville County's three farms produce in 1989? _____

5. In which year did Dayville County produce the fewest pounds of tomatoes? _____

6. Which farm had the most dramatic increase in production from one year to the next? _____

7. How many more pounds of tomatoes did Jefferson's Farm produce in 1990 than in 1986? _____

8. Which farm produced the most pounds of tomatoes in 1993? _____

CIRCLE GRAPHS

Circle graphs represent data expressed in percentages of a total. The parts in a circle graph should always add up to 100%. Circle graphs are sometimes called **pie graphs** or **pie charts**.

To figure the value of a percent in a circle graph, multiply the percent by the total. Use the circle graphs below to answer the questions. The first question is worked for you as an example.

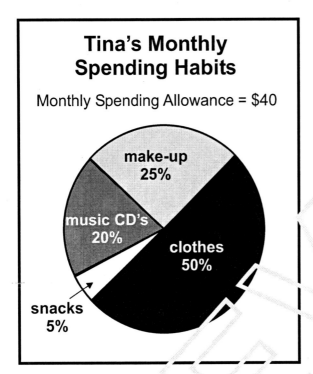

1. How much did Tina spend each month on music CD's?

   $40 × 0.20 = $8.00 $8.00

2. How much did Tina spend each month on make-up?   _____

3. How much did Tina spend each month on clothes?   _____

4. How much did Tina spend each month on snacks?   _____

Fill in the following chart.

Favorite Activity	Number of students
5. watching TV	1000 × 0.3 = 300
6. talking on the phone	
7. playing video games	
8. surfing the Internet	
9. playing sports	
10. reading	

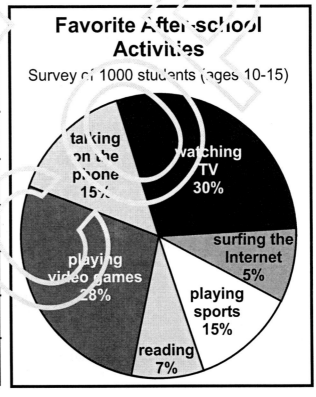

154   Copyright © American Book Company

CATALOG ORDERING

Follow directions for catalog ordering carefully.

DOOGIE'S DOG TREATS
The treats dogs beg for!

Doogie's dog treats are made with only the finest ingredients. They contain real meat and are low in fat with no added sugar or salt.

Flavors:
Cheese   Beef
Lamb & Rice   Peanut Butter

2 lb bag of any 1 flavor ... $4.75
4 lb bag of any 1 flavor ... $6.50
20 lb bag of any 1 flavor ... $25.25

Shipping & Handling
If total order is:

$0 to $10.00 add $3.75
$10.01 to $20.00 add ... $4.25
$20.01 to $30.00 add ... $5.75
$30.01 & Over add $6.25

Figure the total amount to send with each of the following Doogie Dog Treat orders.

1. one 2 lb bag of beef flavored treats

   _____

2. one 20 lb bag of beef treats, two 4 lb bags of cheese treats, and one 2 lb bag of peanut butter treats

   _____

3. two 2 lb bags of lamb & rice treats, two 4 lb bags of beef treats, one 4 lb bag of cheese treats

   _____

4. one 20 lb bag of each flavor

   _____

5. one 4 lb bag of each flavor

   _____

CD BARGAIN CELLAR
Rap · Pop · Country · Rock · Gospel

All "classic" CD's ... $10.95   All new release CD's ... $14.95

Take a 20% discount on all purchases over $50.00!

We have all the hits: Old and New! Order Now!

Figure the total amount to send with each of the following CD Bargain Cellar orders. Don't forget to take discounts when possible.

6. 2 new release rock CD's and 4 "classic" rock CD's   _____

7. 5 "classic" country CD's   _____

8. 1 new release pop CD, 1 new release rap CD, 1 "classic" gospel CD   _____

9. 4 "classic" country CD's and 1 new release country CD   _____

MENU

PANCAKE CAFÉ

CLASSIC BREAKFASTS

Bacon or Sausage Add $1.90   Ham Add $1.80   Hash Browns Add $1.20   Grits Add $.90
Orange Juice Add $2.00   Coffee Add $1.20   Tea Add $1.20

Pancakes		Omelets		Sampler Breakfasts	
Buttermilk (4)	$3.40	Country Omelet	$6.00	2 Eggs, 2 Pancakes, 2 Bacon	$5.50
(2)	$2.50	Vegetable Omelet	$5.50	2 Eggs, Hash Browns, 2 Sausage	$5.50
Pecan	$4.00	Steak Omelet	$6.50	2 Eggs, 2 Biscuits, 2 Bacon	$5.50
Chocolate Chip	$4.00	Denver Omelet	$5.70	Eggs Benedict, Toast, Hash Browns	$5.70
Blueberry	$3.50	Sausage Omelet	$5.60	Steak and Eggs, Toast, Hash Browns	$6.50
Buckwheat	$4.00	Cheese Omelet	$6.20	French Toast, 2 Bacon, 2 Sausage	$5.80

Figure the total cost of the following orders. Round to the nearest penny.

Total Cost

1. Vegetable omelet, hash browns, orange juice, coffee + 5% sales tax   _____

2. Chocolate chip pancakes, orange juice, ham + 6% sales tax   _____

3. Steak omelet, grits, and tea + 7% sales tax   _____

4. 2 eggs, 2 biscuits, 2 bacon, coffee + 5% sales tax   _____

5. French toast, 2 bacon, 2 sausage, grits, coffee + 4% sales tax   _____

6. Denver omelet, grits, hash browns, orange juice + 5% sales tax   _____

7. Buckwheat pancakes, ham, coffee, orange juice + 6% sales tax   _____

8. 2 eggs, 2 pancakes, 2 bacon, orange juice, grits, coffee + 4% sales tax   _____

9. 2 buttermilk pancakes, sausage, coffee + 7% sales tax   _____

10. Eggs Benedict, toast, hash browns, tea + 6% sales tax   _____

11. Cheese omelet, orange juice, bacon, grits, coffee + 5% sales tax   _____

12. Blueberry pancakes, sausage, orange juice, coffee   _____

CHAPTER REVIEW

KNIGHTS BASKETBALL Points Scored				
Player	game 1	game 2	game 3	game 4
Joey	5	2	4	8
Lennard	10	8	10	12
Myron	2	6	5	6
Ned	1	3	6	2
Phil	0	4	7	5
Warren	7	2	9	4
Zeek	8	6	7	4

1. How many points did the Knights basketball team score in game 1?

   _____

2. How many more points did Warren score in game 3 than in game 1?

   _____

3. How many points did Lennard score in the first 4 games?

   _____

Use the following catalogue information to answer questions 4 & 5.

4. Tommy wants to order 1 basic model airplane set and 2 deluxe sets. How much money should he send with his order?

   _____

5. How much would it cost to order 2 standard model airplane sets and 1 deluxe set?

   _____

Model Airplanes
B1 Bombers, F14's, F15's, F16's, and more!

Basic $21.90
(includes only building materials)

Standard $23.90
(includes building materials plus paint)

Deluxe $35.90
(includes building materials, paint, and decals plus a pamphlet on the history of the aircraft)

Add 10% for shipping and handling

HAPPY'S ICE-CREAM SHOP

Scoops
Cup cake cone or sugar cone:
Single $1.10
Double ... $1.75
Triple $2.25
Waffle Cone: Add 50¢
Each topping / Additional toppings: Add 50¢

Sundaes
includes: Ice-cream, 2 toppings, Whipped cream and Cherry
Child $1.50
Small $2.10
Regular $2.90
Large $3.55

Flavors: Chocolate, Vanilla, Strawberry, Chocolate Chip, Tropical, Black Cherry, Peach, Butter Pecan, Banana Nut
Toppings: Hot fudge, Hot caramel, Hot butterscotch, Almonds Chocolate chips, Sprinkles, Wet walnuts, Peanut butter pieces

6. Sheldon ordered a regular sundae with chocolate ice-cream, hot fudge, and almonds. He chose an additional topping of sprinkles. What was the cost of his sundae?

   _____

7. Klarissa ordered a double scoop of chocolate chip ice-cream in a waffle cone with peanut butter pieces. What was the total cost of her order?

   _____

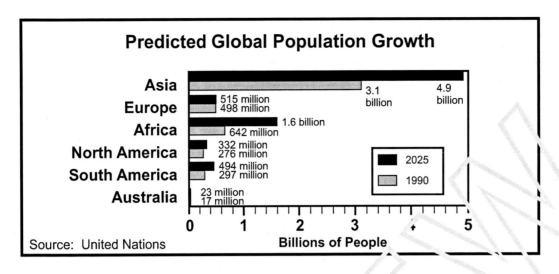

8. Between 1990 and 2025, how much is Asia's population predicted to increase?

9. In 1990, how much larger was Africa's population than Europe's?

10. Between 1990 and 2025, how much is Africa's population predicted to increase?

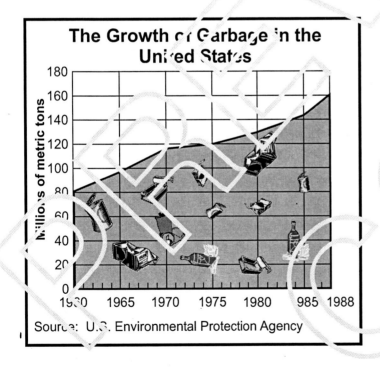

11. How much did the volume of garbage grow between 1960 and 1988?

12. In which year did garbage in the United States reach 140 million metric tons?

13. How much did the volume of garbage grow between 1960 and 1966?

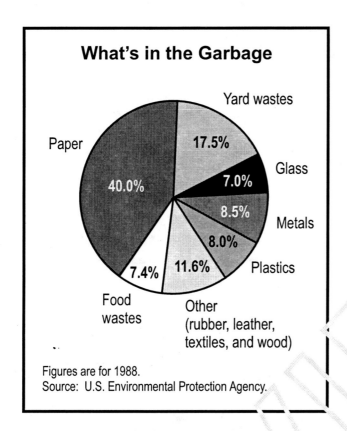

14. In 1988, the United States produced 160 million metric tons of garbage. According to the pie chart, how much glass was in the garbage?

_____

15. Out of the 160 million tons of garbage, how much was glass, plastic, and metal?

_____

16. If in 1990, the garbage reached 200 million metric tons, and the percentage of wastes remained the same as in 1988, how much food would have been in the 1990 garbage?

_____

The students at a middle school raised money for a school-wide field day by selling candy bars for a month. The graph below shows sales per grade each week.

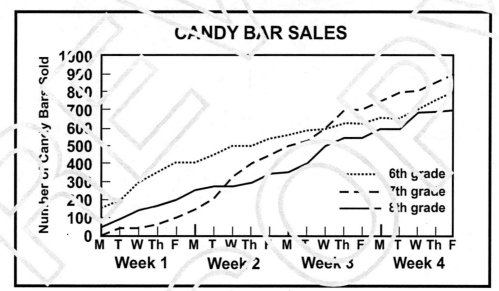

17. On Friday of week 1, how many more candy bars had the 6th grade sold compared to the 7th grade? _____

18. By the end of week 3, which grade had sold the most candy bars? _____

19. On Thursday of week 2, how many candy bars had the 6th, 7th, and 8th grades sold? _____

20. How many total candy bars did all three grades sell by the end of week 4? _____

CHAPTER TEST

1. The Thompson family wants to make a graph to show the percent of money spent on vacation in the following categories:

entertainment	5%
hotel	55%
souvenirs	8%
food	20%
gasoline	12%

   What is the best kind of graph to use to display their results?

   A. bar graph
   B. single line graph
   C. circle graph
   D. multiple line graph

Look at the catalog chart below and answer questions 2 & 3.

SOCKS-A-PLENTY WAREHOUSE		
Item	3 pair price	6 pair price
Anklets	$4.50	$ 8.50
Sport Socks	$6.50	$11.50
Support Socks	$9.00	$15.75
Knee Hi	$6.00	$11.25
Please add 10% for shipping and handling		

2. What would be the total cost of 3 pairs of Knee Hi's, 6 pairs of Anklets, and 3 pairs of Support Socks?

   A. $23.50
   B. $25.85
   C. $33.50
   D. $77.55

3. What would be the total cost of 18 pairs of Sports Socks and 12 pairs of Anklets?

   A. $56.65
   B. $51.50
   C. $57.00
   D. $339.90

4. According to the bar graph below, what year had the greatest increase in the kangaroo population at **Zoo Down Under**?

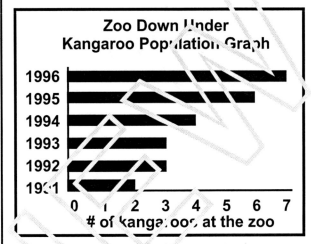

   A. 1996
   B. 1993
   C. 1992
   D. 1994

5. **LANE PUBLISHING COMPANY**
   1996 Total Sales = $1,000,000

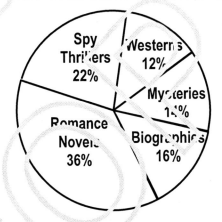

   According to the pie graph, how much of the 1996 total sales came from romance novels?

   A. $120,000
   B. $160,000
   C. $220,000
   D. $360,000

The following chart reflects the number of sheep and goats on a farm.

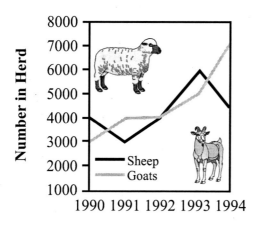

	Jacksonville	Miami	Orlando	Tallahassee	Tampa	West Palm Beach
Jacksonville	0	341	142	163	202	277
Miami	341	0	229	478	273	66
Orlando	142	229	0	257	84	165
Tallahassee	163	478	257	0	275	414
Tampa	202	273	84	275	0	226
West Palm Beach	277	66	165	414	226	0

6. Which year did the sheep outnumber the goats by 1000?

   A. 1991
   B. 1994
   C. 1993
   D. 1994

7. By how many did the number of sheep increase between 1992 and 1993?

   A. 500
   B. 1000
   C. 1500
   D. 2000

Time	Zone 1	Zone 2
7 am to 6 pm	$.09/min	$.12/min
6 pm to 9 pm	$.07/min	$.10/min
9 pm to 7 am	$.05/min	$.07/min
Holidays	$.05/min	$.07/min

8. The table above is a long distance telephone rate chart. How much would a 12 minute call be to Zone 2 from 6:30 pm to 6:42 pm?

   A. $0.99
   B. $1.20
   C. $0.84
   D. $1.12

9. According to the mileage chart above, how far is Tallahassee from Orlando?

   A. 84 miles
   B. 142 miles
   C. 257 miles
   D. 478 miles

10. The Kreise family drove from Miami to West Palm Beach, and then they drove from West Palm Beach to Orlando. How many miles did they travel total?

    A. 66
    B. 165
    C. 202
    D. 231

Brand	Fat Grams	Fat Calories
Yummy	9 g/cup	81 calories
Delite	3 g/cup	27 calories
Kreamy	8 g/cup	72 calories

11. Looking at the chart above, calculate the fat grams for Ben's extra big bowl of ice cream. Ben had 1 cup of Yummy, 2 cups of Delite, and $1\frac{1}{2}$ cups of Kreamy.

    A. 24 grams
    B. 20 grams
    C. 23 grams
    D. 27 grams

Use the following menu to answer question 12.

BARBEQUE

Platters (includes cole-slaw, baked beans, and hot rolls)

	Small	Large	All-you-can-eat
Beef	$4.25	$5.25	$6.75
Pork	$4.75	$5.75	$7.25
Chicken	$4.25	$5.25	$6.75

Side items ($1.25 each): side salad, corn-on-the-cob, or Brunswick stew

Drinks ($0.75): iced tea, cola, coffee

12. Clark ordered an all-you-can-eat beef platter with a side of Brunswick stew and iced tea for lunch. How much did he pay for lunch?

    A. $8.75
    B. $8.00
    C. $7.50
    D. $6.75

Tax Table - Single Person			
Wages		Exemptions	
At least	But less than	0	1
$136.00	$176.00	1.30	$0.18
176.00	216.00	2.25	0.85
216.00	256.00	3.61	1.69
256.00	296.00	5.21	2.89
296.00	335.00	6.97	4.27

13. According to the tax table above, how much total tax should be withheld for the following 3 employees: Dan who earned $272.87 and has 0 exemptions, Sue who earned $212.14 and has 1 exemption, and Diane who earned $157.65 and has 1 exemption.

    A. $6.24
    B. $7.75
    C. $8.00
    D. $8.87

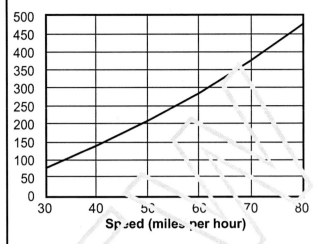

14. According to the chart above, how many more feet does it take to stop a car traveling at 70 miles per hour than at 55 miles per hour?

    A. 175 feet
    B. 150 feet
    C. 125 feet
    D. 100 feet

15. According to the graph below, approximately what percent of people in the USA had a high school diploma in 1994?

    LEVELS OF EDUCATION
    Persons 25 to 34 Years Old
    Data taken from the Statistical Abstract of the United States, 1994

    Key:
    ☐ Percent of people with no high school diploma
    ▨ Percent of people with a high school diploma
    ■ Percent of people with a college degree

    A. 15%
    B. 50%
    C. 65%
    D. 75%

Chapter 11

STATISTICS

Statistics is a branch of mathematics. Using statistics, mathematicians organize data (numbers) into forms that are easily understood.

RANGE

In **statistics,** the difference between the largest number and the smallest number in a list is called the **range.**

EXAMPLE: Find the range of the following list of numbers: 16, 73, 26, 15, and 35.

The largest number is 73, and the smallest number is 15. 73 − 15 = 58
The range is 58.

Find the range for each list of numbers below.

1. 21	2. 6	3. 89	4. 41	5. 23	6. 2	7. 77
51	7	22	3	20	38	91
48	31	65	55	64	29	27
42	55	36	41	38	35	46
12	8	20	19	21	59	63

8. 51	9. 65	10. 84	11. 84	12. 21	13. 45	14. 62
62	54	59	65	78	57	39
32	56	48	32	6	57	96
16	5	21	50	97	14	45
59	63	80	71	45	61	14

15. 2, 13, 3, 25, and 17   range _____

16. 15, 48, 52, 41, and 8   range _____

17. 54, 74, 2, 86, and 75   range _____

18. 15, 61, 11, 22, and 65   range _____

19. 33, 18, 65, 12, and 74   range _____

20. 47, 12, 33, 25, and 19   range _____

21. 56, 10, 33, 7, 16, and 5   range _____

22. 46, 25, 78, 49, and 6   range _____

23. 45, 75, 65, and 21   range _____

24. 97, 23, 56, 12, and 66   range _____

25. 87, 44, 63, and 12   range _____

26. 84, 55, 66, 38, and 31   range _____

27. 35, 44, 81, 99, and 78   range _____

28. 95, 54, 62, 14, 8, and 3   range _____

MEAN

In statistics, the **mean** is the same as the **average**. To find the **mean** of a list of numbers, first add together all the numbers in the list, and then divide by the number of items in the list.

EXAMPLE: Find the mean of: 38, 72, 110, 548

Step 1: First Add: $38 + 72 + 110 + 548 = 768$

Step 2: There are 4 numbers in the list, so divide the total by 4. $\quad 768 \div 4 = 192$
The mean is 192.

Practice finding the mean (average). Round to the nearest tenth if necessary.

1. Dinners served:
   489 561 522 450
   Mean = _____

2. Prices paid for shirts:
   $4.89 $9.97 $5.90 $8.64
   Mean = _____

3. Piglets born:
   23 19 15 21 22
   Mean = _____

4. Student Absences:
   6 5 13 8 9 12 7
   Mean = _____

5. Paychecks:
   $89.56 $99.99 $56.54
   Mean = _____

6. Choir attendance:
   55 45 97 66 70
   Mean = _____

7. Long distance calls:
   33 14 24 21 19
   Mean = _____

8. Train boxcars:
   56 55 48 61 51
   Mean = _____

9. Cookies eaten:
   5 6 8 9 2 4 3
   Mean = _____

Find the mean (average) of the following word problems.

10. Val's science grades were: 85, 87, 65, 94, 78, and 97. What was her average? _____

11. Ann runs a business from her home. The number of orders for the last 7 business days were: 17, 24, 13, 8, 11, 15, and 9. What was the average number of orders per day? _____

12. Melissa tracked the number of phone calls she had per day: 8, 2, 5, 4, 7, 5, 6, 1. What was the average number of calls she received? _____

13. The Cheese Shop tracked the number of lunches they served this week: 42, 55, 36, 41, 38, 33, and 46. What was the average number of lunches served? _____

14. Leah drove 364 miles in 7 hours. What was her average miles per hour? _____

15. Tim saved $680 in 8 months. How much did his savings average each month? _____

16. Ken made 117 passes in 13 games. How many passes did he average per game? _____

MEDIAN

In a list of numbers ordered from lowest to highest, the **median** is the middle number. To find the **median,** first arrange the numbers in numerical order. If there is an odd number of items in the list, the **median** is the middle number. If there is an even number of items in the list, the **median** is the **average of the two middle numbers.**

EXAMPLE 1: Find the median of: 42, 35, 45, 37, and 41.

Step 1: Arrange the numbers in numerical order: 35 37 (41) 42 45

Step 2: Find the middle number. **The median is 41.**

EXAMPLE 2: Find the median of 14, 53, 42, 6, 14, and 46.

Step 1: Arrange the numbers in numerical order: 6 14 (14 42) 46 53

Step 2: Find the average of the 2 middle numbers.
(14 + 42) ÷ 2 = 28. **The median is 28.**

Circle the median in each list of numbers.

1. 35, 55, 40, 30, and 45
2. 7, 2, 3, 6, 5, 1, and 8
3. 65, 42, 60, 46, and 50
4. 15, 16, 19, 25, and 20
5. 75, 90, 87, 65, 82, 88, and 100
6. 33, 42, 30, 22, and 19
7. 401, 758, and 254
8. 41, 23, 14, 21, and 19
9. 5, 8, 3, 10, 13, 1 and 8

10.	11.	12.	13.	14.	15.	16.
19	9	45	52	20	3	15
14	3	32	54	21	17	40
12	10	36	19	25	13	42
15	17	55	63	18	14	32
18	6	61	20	16	22	28

Find the median in each list of numbers.

17. 10, 8, 21, 14, 9 and 12 _____
18. 43, 36, 20, and 40 _____
19. 5, 24, 9, 18, 12 and 3 _____
20. 48, 13, 54, 82, 90, and 7 _____
21. 23, 21, 30, and 27 _____
22. 9, 4, 3, 1, 6, 2, 10, and 12 _____

23.	24.	25.	26.	27.	28.	29.
2	11	13	75	48	22	17
10	22	15	62	45	19	30
6	25	9	60	52	15	31
18	28	35	52	30	43	18
20	10	29	80	35	34	14
23	23	33	50	58	28	25

_____ _____ _____ _____ _____ _____ _____

MODE

In statistics, the **mode** is the number that occurs most frequently in a list of numbers.

EXAMPLE: Exam grades for a Math class were as follows:
70 88 92 85 99 85 70 85 99 100 88 70 99 88 88 99 88 92 85 88

Step 1: Count the number of times each number occurs in the list.

70 - 3 times
88 - 6 times
92 - 2 times
85 - 4 times
99 - 4 times
100 - 1 time

Step 2: Find the number that occurs most often.
The mode is 88 because it is listed 6 times. No other number is listed as often.

Find the mode in each of the following lists of numbers:

1. 88	2. 54	3. 21	4. 56	5. 24	6. 5	7. 12
15	42	16	67	22	4	4
88	44	15	67	22	9	45
17	56	78	19	15	8	32
18	44	21	56	14	4	16
88	44	15	67	14	7	12
17	56	21	20	22	4	12
mode ___	mode ___	mode ___	mode ___	mode ___	mode ___	mode ___

8. 48, 32, 56, 32, 56, 48, 56   mode _____
9. 12, 16, 54, 78, 16, 22, 20   mode _____
10. 5, 4, 8, 3, 4, 2, 7, 8, 4, 2   mode _____
11. 11, 9, 7, 11, 7, 5, 7, 7, 5   mode _____
12. 84, 22, 79, 22, 87, 22, 22   mode _____
13. 92, 87, 65, 94, 78, 95   mode _____
14. 8, 2, 5, 4, 7, 2, 3, 6, 1   mode _____
15. 89, 7, 11, 89, 17, 56   mode _____
16. 15, 48, 52, 41, 8, 48   mode _____

17. 22, 45, 48, 12, 22, 41, 22   mode _____
18. 62, 44, 78, 62, 54, 44, 62   mode _____
19. 54, 22, 54, 78, 22, 78, 22   mode _____
20. 14, 17, 23, 21, 33, 17, 33   mode _____
21. 65, 51, 8, 21, 8, 65, 70, 8   mode _____
22. 17, 24, 3, 8, 11, 8, 15, 9   mode _____
23. 51, 45, 84, 51, 65, 74, 51   mode _____
24. 8, 74, 65, 15, 9, 10, 74   mode _____
25. 62, 54, 2, 7, 89, 2, 7, 54, 2 mode _____

TALLY CHARTS AND FREQUENCY TABLES

Large lists of data can be tallied in a chart. To make a **tally chart**, record a tally mark in a chart for each time a number is repeated. To make a **frequency table**, count the times each number occurs in the list, and record the frequency.

EXAMPLE 1: The age of each student in grades 6-8 in a local middle school are listed below. Make a tally chart and a frequency table for each age.

STUDENT AGES grades 6-8
10, 11, 11, 12, 14, 12,
13, 13, 13, 12, 14, 11,
12, 14, 12, 10, 15, 11,
12, 10, 12, 11, 12, 13,
13, 12, 13, 12, 11, 10,
14, 14, 11, 15, 12, 13,
12, 11, 14, 12, 11, 13,
11, 12, 13, 12, 13, 14

TALLY CHART	
Age	Tally
10	IIII
11	HHH HHH
12	HHH HHH HHH
13	HHH HHH
14	HHH II
15	II

FREQUENCY TABLE	
Age	Frequency
10	4
11	10
12	15
13	10
14	7
15	2

Sometimes, a large list of data can be best understood if the numbers are grouped in intervals and tallied according to those intervals.

EXAMPLE 2: A local drug store wants to make a tally chart/frequency table to show the total purchase amount per customer grouped in intervals of $5.00. The following list of data gives the sales totals for each customer for one day.

$34.75	$10.42	$12.34	$17.63	$37.04	$15.68	$18.42	$19.75
$19.65	$8.32	$5.02	$23.08	$12.42	$5.67	$25.45	$19.49
$7.82	$2.25	$1.96	$34.54	$18.52	$17.28	$5.74	$18.47
$20.57	$10.35	$32.58	$18.50	$6.05	$19.54	$4.57	$32.58
$11.90	$5.57	$7.24	$2.56	$8.97	$27.56	$18.97	$13.00
$9.54	$10.75	$15.32	$20.56	$15.95	$21.02	$11.21	$5.64
$10.08	$14.10	$4.57	$12.17	$19.65	$2.01	$12.07	$27.65

In this case, we will show the tallies and frequencies in one chart.

CUSTOMER SALES		
Total	Tally	Frequency
$0-5.00	HHH I	6
$5.01-10.00	HHH HHH I	11
$10.01-15.00	HHH HHH II	12
$15.01-20.00	HHH HHH HHH	15
$20.01-25.00	III	3
$25.01-30.00	IIII	4
$30.01-35.00	IIII	4
$35.01-40.00	I	1

Make a chart to record tallies and frequencies for the following problems.

1. Mr. Kyle gave all his students a 10 question pop quiz. The following list gives the question numbers answered incorrectly by his students. Fill in the tally and frequency chart.

Question Numbers Missed							
3	4	2	1	3	4	1	6
8	5	8	9	10	4	3	2
1	2	3	4	5	8	9	4
7	4	7	5	3	2	4	5
2	4	3	8	4	6	4	3
8	2	7	5	3	4	3	8
7	8	2	1	7	10	6	5
10	5	2	3	6	4	3	1

Question Number	Tally	Frequency
1		
2		
3		
4		
5		
6		
7		
8		
9		
10		

2. The sheriff's office monitored the speed of cars traveling on Turner Road for one week. The following data is the speed of each car that traveled Turner Road during the week. Tally the data in 10 miles per hour (mph) increments starting with 0-9 mph and record the frequency in a chart.

CAR SPEED, mph							
45	52	47	35	48	50	51	43
40	51	32	24	55	41	32	33
36	39	49	52	34	28	39	47
29	15	63	42	35	42	58	59
39	41	25	34	22	16	40	31
55	10	46	38	30	52	58	56
21	32	36	41	52	49	45	32
52	45	56	35	55	65	20	41

Speed	Tally	Frequency
0-9		
10-19		
20-29		
30-39		
40-49		
50-59		
60-69		

3. The following data gives final math averages for Ms. Kirby's class. In her class, an average of 90-100 is an A, 80-89 is a B, 70-79 is a C, 60-69 is a D, and an average below 60 is an F. Tally and record the frequency of A's, B's, C's, D's, and F's.

Final Math Averages						
85	92	87	62	75	84	96	52
45	7	98	75	71	79	85	82
87	74	76	68	93	77	65	84
79	65	77	82	86	84	92	60
99	75	88	74	79	80	63	84
87	90	75	81	73	69	73	75
31	86	89	65	69	75	79	76

Grade	Tally	Frequency
A		
B		
C		
D		
F		

HISTOGRAMS

A **histogram** is a bar graph of the data in a frequency table.

EXAMPLE: Draw a histogram for the customer sales data presented in the frequency table.

CUSTOMER SALES	
Total	Frequency
$0-5.00	6
$5.01-10.00	11
$10.01-15.00	12
$15.01-20.00	15
$20.01-25.00	3
$25.01-30.00	4
$30.01-35.00	4
$35.01-40.00	1

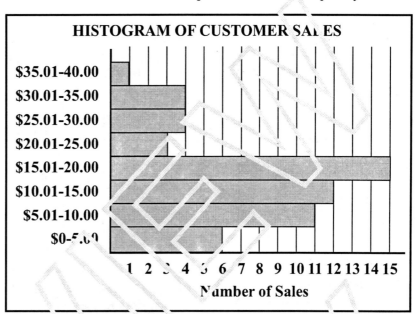

Use the frequency charts that you filled in on the previous page to draw histograms for the same data.

1.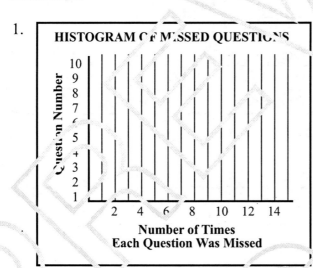

2.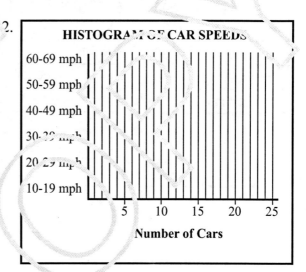

3.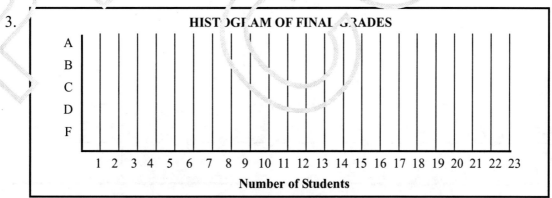

CHAPTER REVIEW

Find the mean, median, mode, and range for each of the following sets of data. Fill in the table below.

❶ Miles Run by Track Team Members

Jeff	24
Eric	20
Craig	19
Simon	20
Elijah	25
Rich	19
Marcus	20

❷ 1992 SUMMER OLYMPIC GAMES Gold Medals Won

Unified Team	45	Hungary	11
United States	37	South Korea	12
Germany	33	France	8
China	16	Australia	7
Cuba	14	Japan	3
Spain	13		

❸ Hardware Store Payroll June Week 2

Erica	$280
Dave	$206
Sam	$240
Nancy	$404
Elsie	$210
Gail	$305
David	$280

❹ Lunches Served

Fairfield's	80
Mikey's	106
House of China	54
Anticoli's	105
Rib Ranch	83
Two Brothers	66
Mountain Inn	80

❺ Average Days of Rain or Snow BIRMINGHAM

January	12	July	12
February	10	August	11
March	11	September	8
April	11	October	6
May	9	November	8
June	11	December	11

❻ PHONE BILLS

January	$80.50
February	$71.97
March	$82.02
April	$80.50
May	$98.19
June	$108.82

Data Set Number	Mean	Median	Mode	Range
❶				
❷				
❸				
❹				
❺				
❻				

Use the data given to answer the questions that follow.

The 6th grade did a survey on the number of pets each student had at home. The following gives the data produced by the survey.

NUMBER OF PETS PER STUDENT
0 2 6 2 1 0 4 2 3 3 0 2 3 5 1 4 2 0 5 2 3 3 4 3 6 2
5 1 2 3 5 6 3 2 2 5 2 3 4 3 0 1 4 1 2 4 5 7 ? 1 4 7

1. Fill in the frequency table.   2. Fill in the histogram

Number of Pets	Frequency
0	5
1	6
2	12
3	10
4	7
5	6
6	3
7	2

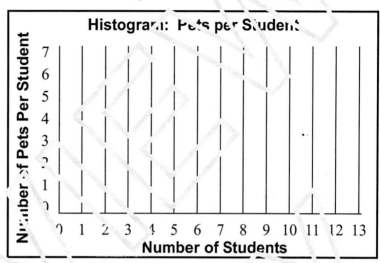

3. What number of pets per student is the mode number? _____

The school store manager kept a record of how much money the store made each day for a month:

DAILY SCHOOL STORE SALES
$6.10 $2.00 $4.05 $4.25 $5.25 $7.75 $1.60 $2.10 $8.50 $6.15
$5.95 $4.50 $4.65 $3.90 $4.75 $3.60 $5.25 $6.20 $7.20 $5.75

4. Fill in the following tally chart in intervals of one dollar.   5. Fill in the histogram using the same intervals as in #4.

SALES	TALLY
$0-1.00	0
$1-1.99	1
$2-2.99	2
$3-3.99	2
$4-4.99	5
$5-5.99	4
$6-6.99	3
$7-7.99	2
$8-8.99	1

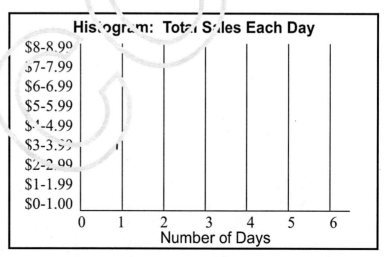

CHAPTER TEST

1. Tom's math grades so far this semester have been 94, 72, 77, 76, and 41. What is the range of his grades?

   A. 53
   B. 72
   C. 76
   D. 94

2. What is the mode of the following set of numbers?

   12, 17, 12, 23, 37, 12, 11, 17

   A. 19
   B. 26
   C. 15
   D. 12

3. **DOBBINS FAMILY**

Month	Electric Bill
January	$89.15
February	$99.59
March	$73.90
April	$72.47
May	$99.23
June	$124.69

   What is the median electric bill for the Dobbins family from January through June?

   A. $ 55.22
   B. $ 94.02
   C. $ 94.19
   D. $124.69

4. Concession stand sales for each game in the season were:
   $350, $540, $230, $450, $280, $580

   What is mean sales per game?

   A. $230
   B. $350
   C. $385
   D. $400

5. The band teacher took a survey of reasons students took his class. He made the following histogram from the data he collected.

   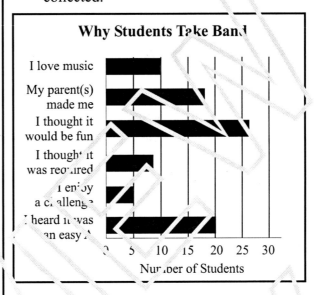

   Which reason for taking band is the mode of the data?

   A. I enjoy a challenge
   B. I love music
   C. I thought is would be fun
   D. I heard it was an easy A

6. Rachel kept track of how many scoops she sold of the five most popular flavors in her ice cream shop.

Flavor	Scoops
Vanilla Bean	30
Chunky Chocolate	36
Strawberry Coconut	44
Chocolate Peanut Butter	46
Mint Chocolate Chip	28

   Which flavor of ice cream is closest to the mean number?

   A. Vanilla Bean
   B. Chocolate Peanut Butter
   C. Chunky Chocolate
   D. Strawberry Coconut

7. Below is a list of John and Mary's children. What is the range of the children's ages?

   Harold   17 years old
   Terry   15 years old
   Brenda   12 years old
   Alex   9 years old
   Colby   4 years old

   A. 13
   B. 12
   C. 11
   D. 5

8. Which set of numbers has the greatest range?

   A. { 95, 86, 78, 62 }
   B. { 90, 65, 83, 59 }
   C. { 32, 29, 44, 56 }
   D. { 29, 35, 49, 51 }

9. Zena earned $30 for each of the last 3 weeks. This week she earned $42. What were her average earnings for the 4 weeks?

   A. $30.00
   B. $33.00
   C. $36.00
   D. $42.00

10. What is the median of 27, 32, 16, 31, 9 and 37?

    A. 31
    B. 16
    C. 28
    D. 29

11. What is the mean of 12, 23, 8, 26, 37, 11, and 9?

    A. 12
    B. 29
    C. 18
    D. 19

12. The student council surveyed the student body on favorite lunch items. The frequency chart below shows the results of the survey.

Favorite Lunch Item	Frequency
corndog	140
hamburger	245
hotdog	210
pizza	235
spaghetti	90
other	65

    Which lunch item indicates the mode of the data?

    A. other
    B. hotdog
    C. hamburger
    D. corndog

13. Which of the following sets of numbers has a median of 42?

    A. { 60, 42, 37, 22, 19 }
    B. { 16, 28, 42, 48 }
    C. { 42, 64, 20 }
    D. { 12, 42, 40, 50 }

14. Which of the following sets of numbers has a range of 51?

    A. { 22, 19, 72, 68, 59 }
    B. { 81, 35, 37, 41, 60 }
    C. { 17, 12, 9, 47, 82 }
    D. { 62, 36, 44, 78, 95 }

15. Find the mode of 7, 12, 15, 7, 4, 1, and 10.

    A. 7
    B. 8
    C. 1
    D. 6

16. A restaurant that serves breakfast between 7:00 a.m. and 11:00 a.m. recorded the times of the morning each customer was served. The histogram below gives data for a typical week.

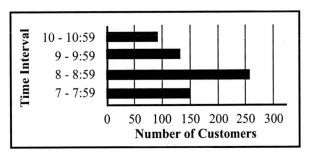

Which interval represents the mode of the data?

A. 7 - 7:59
B. 8 - 8:59
C. 9 - 9:59
D. 10 - 10:59

17. What is the range of 12, 73, 82, 17, and 46?

A. 5
B. 46
C. 70
D. 72

18. The following list represents the prices of athletic shoes sold one day at a local retail store rounded to the nearest dollar:

$57, $35, $72, $51, $43, $18, $64, $58, $27, $31, $52, $22, $31, $19, $29, $68

If the store manager made a frequency chart of the data in increments of $10, how many shoe sales would be in the $20 - $29 range?

A. 1
B. 2
C. 3
D. 4

19. A neighborhood surveyed the times of day people water their lawns and tallied the data below.

Time	Tally
midnight - 3:59 a.m.	II
4:00 a.m. - 7:59 a.m.	⊬H I
8:00 a.m. - 11:59 a.m.	⊬H IIII
noon - 3:59 p.m.	⊬H
4:00 p.m. - 7:59 p.m.	⊬H ⊬H
8:00 p.m. - 11:59 p.m.	⊬H III

Which time of day is the most popular for the people in this neighborhood to water their lawns?

A. 4:00 a.m. - 7:59 a.m.
B. 8:00 a.m. - 11:59 a.m.
C. 4:00 p.m. - 7:59 p.m.
D. 8:00 p.m. - 11:59 p.m.

20. The following are student run times for the 50 meter dash:

Nate: 3 sec, Pete: 4 sec, Ron: 4 sec, Kate: 5 sec, Wanda: 6 sec, Adam: 7 sec, Jason: 7 sec, May: 8 sec, and Cy: 8 sec

Which student represents the median time?

A. Nate
B. Kate
C. Wanda
D. Cy

21. Gail's science grades were 80, 89, 92, and 71. What was the mean of her grades?

A. 21
B. 33
C. 85
D. 71

Chapter 12

INTEGERS AND ORDER OF OPERATIONS

In elementary school, you learned to use whole numbers.

Whole numbers = { 0, 1, 2, 3, 4, 5 . . . }

For most things in life, whole numbers are all we need to use. However, when your checking account falls below zero or the temperature falls below zero, we need a way to express that. Mathematicians have decided that a negative sign, which looks exactly like a subtraction sign, would be used in front of a number to show that the number is below zero. All the negative whole numbers and positive whole numbers plus zero make up the set of integers.

Integers = { . . . –4, –3, –2, –1, 0, 1, 2, 3, 4 . . . }

For pictures at the right, tell how far each object is above or below sea level by using positive (+) and negative (–) numbers on the number line.

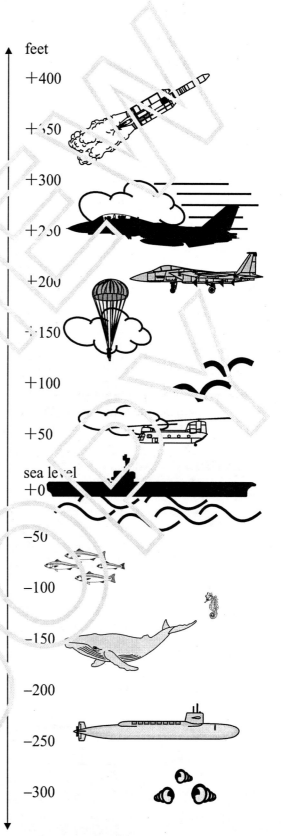

1. Nuclear Submarine _____
2. F-14A (black) _____
3. school of fish _____
4. Titan Rocket _____
5. whale _____
6. F-15 (gray) _____
7. seagulls _____
8. shells _____
9. Chinook Helicopter _____
10. What if the whale went up 50 feet (+50)? Where would it be? _____
11. What if the submarine went down 50 feet (–50)? Where would it be? _____
12. What if the seahorse went up 20 feet? Where would it be then? _____
13. What if the whale swam down 25 feet (–25)? Where would it be then? _____

ABSOLUTE VALUE

The absolute value of a number is the distance the number is from zero on the number line.

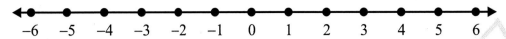

The absolute value of 6 is written | 6 |. | 6 | = 6
The absolute value of −6 is written | −6 |. | −6 | = 6

Both 6 and −6 are the same distance, 6 spaces, from zero so their absolute value is the same, 6.

EXAMPLES:

| −4 | = 4   − | −4 | = −4 | 9 | + | 5 | = 9 + 5 = 14
| 9 | − | 8 | = 9 − 8 = 1 | 6 | − | −6 | = 6 − 6 = 0 | −5 | + | −2 | = 5 + 2 = 7

Simplify the following absolute value problems.

1. | 9 | = _____
2. − | 5 | = _____
3. | −25 | = _____
4. − | −12 | = _____
5. − | 64 | = _____

6. | −4 | − | 3 | = _____
7. | −8 | − | −4 | = _____
8. | 5 | + | −4 | = _____
9. | −2 | + | 6 | = _____
10. | 10 | + | 8 | = _____

11. | 7 | − | 5 | = _____
12. | −12 | + | −2 | = _____
13. | 15 | − | 4 | = _____
14. | −6 | + | 8 | = _____
15. | 17 | − | −5 | = _____

ADDING INTEGERS

First, we will see how to add integers on the number line; then we will learn rules for working the problems without using a number line.

EXAMPLE 1: Add: (−3) + 7

Step 1 The first integer in the problem tells us where to start.
Find the first integer, −3, on the number line.

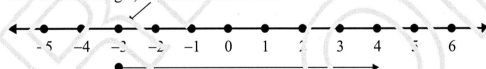

Step 2 (−3) + 7 The second integer in the problem, +7, tells us the direction to go, positive (toward positive numbers), and how far, 7 places. (−3) + 7 = 4

EXAMPLE 2: Add: (−2) + (−3)

Step 1 Find the first integer, (−2), on the number line.

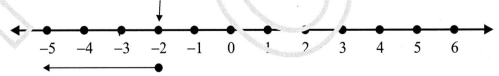

Step 2 (−2) + (−3) The second integer in the problem, (−3), tells us the direction to go, negative (toward the negative numbers), and how far, 3 places. (−2) + (−3) = (−5)

Solve the problems below using this number line.

$\leftarrow \bullet\ \bullet\ \bullet\ \bullet\ \bullet\ \bullet\ \bullet\ \bullet\ \bullet\ \bullet\ \bullet\ \bullet\ \bullet\ \bullet\ \bullet\ \bullet\ \bullet \rightarrow$
$-8\ -7\ -6\ -5\ -4\ -3\ -2\ -1\ \ 0\ \ 1\ \ 2\ \ 3\ \ 4\ \ 5\ \ 6\ \ 7\ \ 8$

1. $2 + (-3) =$ _____
2. $4 + (-2) =$ _____
3. $(-3) + 7 =$ _____
4. $(-4) + 4 =$ _____
5. $(-1) + 5 =$ _____
6. $(-1) + (-4) =$ _____
7. $3 + 2 =$ _____
8. $(-5) + 8 =$ _____

9. $3 + (-7) =$ _____
10. $(-2) + (-2) =$ _____
11. $6 + (-7) =$ _____
12. $2 + (-5) =$ _____
13. $(-5) + 3 =$ _____
14. $(-6) + 7 =$ _____
15. $(-3) + (-2) =$ _____
16. $(-8) + 6 =$ _____

17. $(-2) + 6 =$ _____
18. $(-4) + 8 =$ _____
19. $(-7) + 4 =$ _____
20. $(-5) + 8 =$ _____
21. $(-2) + (-2) =$ _____
22. $3 + (-6) =$ _____
23. $5 + (-3) =$ _____
24. $1 + (-8) =$ _____

RULES FOR ADDING INTEGERS WITH THE SAME SIGNS

To add integers without using the number line, use these simple rules:

> 1. Add the numbers together.
> 2. Give the answer the same sign.

EXAMPLE 1: $(-2) + (-5) =$ _____ Both integers are negative. To find the answer, add the numbers together $(2 + 5)$, and give the answer a negative sign.

$(-2) + (-5) = (-7)$

EXAMPLE 2: $3 + 4 =$ _____ Both integers are positive so the answer is positive.

$3 + 4 = 7$

NOTE: Sometimes positive signs are placed in front of positive numbers. For example $3 + 4 = 7$ may be written $(+3) + (+4) = +7$. Positive signs in front of positive numbers are optional. If a number has no sign, it is considered positive.

Solve the problems below using the rules for adding integers with the same signs.

1. $(-18) + (-4) =$ _____
2. $(-12) + (-3) =$ _____
3. $(-2) + (-7) =$ _____
4. $(+22) + (+11) =$ _____
5. $(-7) + (-6) =$ _____

6. $(-9) + (-8) =$ _____
7. $8 + 4 =$ _____
8. $(-4) + (-7) =$ _____
9. $(-15) + (-5) =$ _____
10. $(+7) + (+4) =$ _____

11. $(-7) + (-20) =$ _____
12. $(-18) + (-16) =$ _____
13. $25 + 32 =$ _____
14. $(-15) + (-3) =$ _____
15. $(9) + (9) =$ _____

RULES FOR ADDING INTEGERS WITH OPPOSITE SIGNS

> 1. Ignore the signs and find the difference.
> 2. Give the answer the sign of the larger number.

EXAMPLE 1: $(-4) + 6 =$ _____

$(-4) + 6 = 2$

To find the difference, take the larger number minus the smaller number. $6 - 4 = 2$. Looking back at the original problem, the larger number, 6, is positive, so the answer is positive.

EXAMPLE 2: $3 + (-7) =$ _____

$3 + (-7) = (-4)$

Find the difference. $7 - 3 = 4$. Looking at the problem, the larger number, 7, is a negative number, so the answer is negative.

Solve the problems below using the rules of adding integers with opposite signs.

1. $(-4) + 8 =$ _____
2. $-10 + 12 =$ _____
3. $9 + (-3) =$ _____
4. $(+3) + (-3) =$ _____
5. $5 + (-2) =$ _____
6. $(-18) + 9 =$ _____
7. $25 + (-30) =$ _____

8. $+8 + (-7) =$ _____
9. $(-5) + (+12) =$ _____
10. $-14 + (+7) =$ _____
11. $7 + (-8) =$ _____
12. $(-30) + 15 =$ _____
13. $100 + (-65) =$ _____
14. $85 + (-14) =$ _____

15. $6 + (-12) =$ _____
16. $(-11) + 1 =$ _____
17. $3 + (-13) =$ _____
18. $(-12) + 8 =$ _____
19. $52 + (-9) =$ _____
20. $(-39) + 8 =$ _____
21. $(+14) + (-16) =$ _____

Solve the mixed addition problems below using the rules for adding integers.

22. $-7 + 8 =$ _____
23. $5 + 6 =$ _____
24. $(-2) + (-6) =$ _____
25. $3 + (-5) =$ _____
26. $(-7) + (-9) =$ _____
27. $14 + 9 =$ _____
28. $(-15) + 6 =$ _____

29. $8 + (-5) =$ _____
30. $(-6) + 13 =$ _____
31. $(-9) + (-12) =$ _____
32. $(-7) + (+12) =$ _____
33. $+8 + (-9) =$ _____
34. $(-13) + (-18) =$ _____
35. $46 + (-52) =$ _____

36. $(-7) + (-10) =$ _____
37. $(+4) + 1 =$ _____
38. $11 + 6 =$ _____
39. $-4 + (-10) =$ _____
40. $(+6) + (+2) =$ _____
41. $1 + (-17) =$ _____
42. $(-42) + 6 =$ _____

RULES FOR SUBTRACTING INTEGERS

To subtract integers, the easiest way is to change the problem to an addition problem and follow the rules you already know.

> 1. Change the subtraction sign to addition.
> 2. Change the sign of the second number to the opposite sign.

EXAMPLE 1: $-6 - (-2) =$ _____ Change the subtraction sign to addition and -2 to 2. $\quad -6 - (-2) = (-6) + 2$

$(-6) + 2 = (-4)$

EXAMPLE 2: $5 - 6 =$ _____ Change the subtraction sign to addition and 6 to -6. $\quad 5 - 6 = 5 + (-6)$

$5 + (-6) = (-1)$

Solve the problems using the rules above.

1. $(-3) - 8 =$ _____
2. $5 - (-9) =$ _____
3. $8 - (-5) =$ _____
4. $(-2) - (-6) =$ _____
5. $8 - (-9) =$ _____
6. $(-4) - (-1) =$ _____

7. $(-5) - (-12) =$ _____
8. $6 - (-7) =$ _____
9. $8 - (-6) =$ _____
10. $(-2) - (-2) =$ _____
11. $(-3) - 7 =$ _____
12. $(-4) - 8 =$ _____

13. $(-7) - 4 =$ _____
14. $1 - (-9) =$ _____
15. $(-5) - 12 =$ _____
16. $(-1) - 9 =$ _____
17. $6 - (-7) =$ _____
18. $(-8) - (-12) =$ _____

Solve the addition and subtraction problems below.

19. $4 - (-2) =$ _____
20. $(-2) + 7 =$ _____
21. $(-4) + 14 =$ _____
22. $(-1) - 5 =$ _____
23. $(-1) + (-4) =$ _____
24. $(-12) + (-2) =$ _____
25. $0 - (-6) =$ _____
26. $2 - (-5) =$ _____

27. $(-5) + 3 =$ _____
28. $(-6) + 7 =$ _____
29. $(-4) + 3 =$ _____
30. $(-4) - 11 =$ _____
31. $(-5) + 8 =$ _____
32. $(-3) - (-3) =$ _____
33. $(-8) + 9 =$ _____
34. $0 + (-10) =$ _____

35. $30 + (-15) =$ _____
36. $-40 - (-5) =$ _____
37. $25 - 50 =$ _____
38. $-13 + 12 =$ _____
39. $(-21) - (-1) =$ _____
40. $62 - (-3) =$ _____
41. $(-16) + (-2) =$ _____
42. $(-25) + 5 =$ _____

MULTIPLYING INTEGERS

You are probably used to seeing multiplication written with a "×" sign, but multiplication can be written two other ways. A "·" between numbers means the same as "×", and parentheses ( ) around a number without a "×" or a "·" also means to multiply.

EXAMPLES: $2 \times 3 = 6$ or $2 \cdot 3 = 6$ or $(2)(3) = 6$

All of these mean the same thing.

DIVIDING INTEGERS

Division is commonly indicated two ways: with a "÷" or in the form of a fraction.

EXAMPLE: $6 \div 3 = 2$ means the same thing as $\frac{6}{3} = 2$

RULES FOR MULTIPLYING AND DIVIDING INTEGERS

1. If the numbers have the same sign, the answer is positive.
2. If the numbers have different signs, the answer is negative.

EXAMPLES: $6 \times 8 = 48$ $(-6) \times 8 = (-48)$ $(-6) \times (-8) = 48$

$48 \div 6 = 8$ $(-48) \div 6 = (-8)$ $(-48) \div (-6) = 8$

Solve the problems below using the rules for dividing integers with opposite signs.

1. $(-4) \div 2 =$ _____
2. $12 \div (-2) =$ _____
3. $\frac{(-14)}{(-2)} =$ _____
4. $-16 \div 5 =$ _____
5. $(-3) \times (-7) =$ _____
6. $(-1) \cdot (5) =$ _____
7. $-1 \times (-4) =$ _____
8. $(3)(2) =$ _____

9. $2(-5) =$ _____
10. $3 \times (-7) =$ _____
11. $(-12) \cdot (-2) =$ _____
12. $\frac{(-18)}{(-6)} =$ _____
13. $21 \div (-7) =$ _____
14. $-5 \times 3 =$ _____
15. $(-6)(7) =$ _____
16. $\frac{(-3)}{(-3)} =$ _____

17. $(-5) \times 8 =$ _____
18. $\frac{-12}{6} =$ _____
19. $8(-4) =$ _____
20. $1 \cdot (-8) =$ _____
21. $(-7) \cdot (-4) =$ _____
22. $(-2) \div (-2) =$ _____
23. $\frac{18}{(-6)} =$ _____
24. $5(-3) =$ _____

MIXED INTEGER PRACTICE

1. $(-6) + 13 =$ _____
2. $(-3) + (-9) =$ _____
3. $(-4) \times 4 =$ _____
4. $(-18) \div 3 =$ _____
5. $(-1) - 5 =$ _____
6. $(-1) \times (-4) =$ _____
7. $3 + (-5) =$ _____
8. $6 + (-5) =$ _____
9. $(-9) - (-12) =$ _____
10. $2 + (-5) =$ _____
11. $\frac{(-24)}{(-6)} =$ _____
12. $(-5) + 3 =$ _____
13. $(-6) - 7 =$ _____
14. $(-33) \div (-11) =$ _____
15. $(-21)(-3) =$ _____
16. $(-7) + (-14) =$ _____
17. $(-5) - 8 =$ _____
18. $1 - (-6) =$ _____
19. $(-2) \cdot (-2) =$ _____
20. $8 + (-6) =$ _____
21. $\frac{-14}{7} =$ _____
22. $(+7) \cdot (-2) =$ _____
23. $(10)(4) =$ _____
24. $24 \div (-4) =$ _____
25. $6(-5) =$ _____
26. $\frac{12}{(-3)} =$ _____
27. $36 \div 12 =$ _____

PROPERTIES OF ADDITION AND MULTIPLICATION

The Associative, Commutative, and Distributive properties and the identity properties of addition and multiplication are listed below by example as a quick refresher.

Property — **Example**

❶ Associative Property of Addition — $(a + b) + c = a + (b + c)$
❷ Associative Property of Multiplication — $(a \times b) \times c = a \times (b \times c)$
❸ Commutative Property of Addition — $a + b = b + a$
❹ Commutative Property of Multiplication — $a \times b = b \times a$
❺ Distributive Property — $a \times (b + c) = (a \times b) + (a \times c)$
❻ Identity Property of Addition — $0 + a = a$
❼ Identity Property of Multiplication — $1 \times a = a$
❽ Inverse Property of Addition — $a + (-a) = 0$
❾ Inverse Property of Multiplication — $a \times \frac{1}{a} = \frac{a}{a} = \quad a \neq 0$

In the blanks provided, write the number of the property listed above that describes each of the following statements.

1. $4 + 5 = 5 + 4$ _____
2. $4 + (2 + 8) = (4 + 2) + 8$ _____
3. $10(4 + 7) = (10)(4) + (10)(7)$ _____
4. $(2 \times 3) \times 4 = 2 \times (3 \times 4)$ _____
5. $1 \times 12 = 12$ _____
6. $8(\frac{1}{8}) = 1$ _____
7. $1c = c$ _____
8. $18 + 0 = 18$ _____
9. $9 + (-9) = 0$ _____
10. $p \times q = q \times p$ _____
11. $t + 0 = t$ _____
12. $x(y + z) = xy + xz$ _____
13. $(m)(n \cdot p) = (m \cdot n)(p)$ _____
14. $-y + y = 0$ _____

ORDER OF OPERATIONS

In long math problems with $+$, $-$, $\times$, $\div$, ( ), and exponents in them, you have to know what to do first. Without following the same rules, you could get different answers. If you will memorize the silly sentence, Please Excuse My Dear Aunt Sally, you can easily memorize the order you must follow.

Please — "P" stands for parentheses. You must get rid of parentheses first.
Examples: $3(1+4) = 3 \times 5 = 15$
$6(10-6) = 6 \times 4 = 24$

Excuse — "E" stands for exponents. You must eliminate exponents next.
Example: $4^2 = 4 \times 4 = 16$

My Dear — "M" stands for multiply. "D" stands for divide. Start on the left of the equation, and perform all multiplications and divisions in the order in which they appear.

Aunt Sally — "A" stands for add. "S" stands for subtract. Start on the left, and perform all additions and subtractions in the order they appear.

EXAMPLE: $12 \div 2(6-3) + 3^2 - 1$

Please	Eliminate **parentheses**. $6 - 3 = 3$ so now we have	$12 \div 2 \times 3 + 3^2 - 1$
Excuse	Eliminate **exponents**. $3^2 = 9$ so now we have	$12 \div 2 \times 3 + 9 - 1$
My Dear	**Multiply** and **divide** next in order from left to right.	$12 \div 2 = 6$ then $6 \times 3 = 18$
Aunt Sally	Last, we **add** and **subtract** in order from left to right.	$18 + 9 - 1 = 26$

Simplify the following problems.

1. $6 + 9 \times 2 - 4 =$ _____
2. $3(4 + 2) - 6^2 =$ _____
3. $3(6 - 3) - 2^2 =$ _____
4. $49 \div 7 - 3 \times 3 =$ _____
5. $10 \times 2 - (7 - 2) =$ _____
6. $2 \times 3 \div 6 \times 4 =$ _____
7. $4 - 8(4 + 2) =$ _____
8. $7 + 8(14 - 6) \div 4 =$ _____
9. $(2 + 8 - 12) \times 4 =$ _____
10. $4(8 - 13) \times 4 =$ _____
11. $8 + 4^2 \times 2 - 6 =$ _____
12. $3^2(4 + 6) + 5 =$ _____
13. $(12 - 6) + 27 \div 3^2 =$ _____
14. $82^0 - 1 + 4 \div 2^2 =$ _____
15. $1 - (2 - 3) + 8 =$ _____
16. $12 - 3(7 - 2) =$ _____
17. $18 \div (6 + 3) - 12 =$ _____
18. $10^2 + 3^3 - 2 \times 3 =$ _____
19. $4^2 + (7 + 2) \div 3 =$ _____
20. $7 \times 4 - 9 \div 3 =$ _____

CHAPTER REVIEW

Simplify the following problems.

1. $(-9) \times (-10) =$ _____
2. $12 + (-22) =$ _____
3. $10 - (-13) =$ _____
4. $12 \div (-3) =$ _____
5. $|-5| + |4| =$ _____
6. $-6 - 5 =$ _____
7. $(-7) \cdot (6) =$ _____
8. $-|9| - |-2| =$ _____
9. $4 - 9 =$ _____
10. $\dfrac{(-25)}{(-5)} =$ _____
11. $(-13) + (-4) =$ _____
12. $(-10)(-6) =$ _____
13. $|-10| + |-2| =$ _____
14. $(-4)(-22) =$ _____
15. $3 + (-9) =$ _____
16. $\dfrac{(-16)}{4} =$ _____
17. $(6)(-11) =$ _____
18. $-4 + (-10) =$ _____
19. $(-24) \div (-6) =$ _____
20. $13 - 18 =$ _____
21. $14 - (-20) =$ _____
22. $\dfrac{45}{(-9)} =$ _____
23. $|-7| + |-5| =$ _____
24. $-3 - (-3) =$ _____
25. $(-1) \times 12 =$ _____
26. $-7 + (-27) =$ _____
27. $9 - (-4) =$ _____
28. $(12)(-3) =$ _____
29. $\dfrac{(-60)}{(-12)} =$ _____
30. $13 - 27 =$ _____
31. $(-15) \times (-2) =$ _____
32. $-|4| - |-8| =$ _____
33. $(-19) + 8 =$ _____
34. $-13 - (-13) =$ _____
35. $(-7) \cdot (-8) =$ _____
36. $36 \div (-6) =$ _____

Simplify the following problems using the correct order of operations.

37. $2^3 \div (2)(8 - 12) =$ _____

38. $20 \div (-2 - 8) + 2 =$ _____

39. $14 + (7)(5 - 5) \div 2 =$ _____

40. $8 - 7^2 + (3 - 12) =$ _____

41. $(18 - 5) \times (2 - 3) - 10 =$ _____

42. $24 - (6^2 - 6) \div 5 =$ _____

43. $3^3 + (7)(9 - 5) =$ _____

44. $-10(1 - 9) \div (-20) + 1 =$ _____

45. $42 \div (12 - 5) - 2 =$ _____

46. $2 + 5^2 \div (15 - 20) =$ _____

CHAPTER TEST

1. $(-17) + (+6) =$

   A. -23
   B. -11
   C. 11
   D. 23

2. $10 \times (-4) =$

   A. $-10{,}000$
   B. -40
   C. -14
   D. $+40$

3. $(749) \div (-7) =$

   A. -17
   B. -107
   C. 17
   D. 107

4. $(-9) - (12) =$

   A. -21
   B. -3
   C. $+3$
   D. $+21$

5. Which of the following is an example of the <u>Distributive Property</u>?

   A. $2 + 11 = 11 + 2$
   B. $7(4 + 3) = (7 \times 4) + (7 \times 3)$
   C. $(5 \times 3) \times 4 = 5 \times (3 \times 4)$
   D. $9 + (1 + 5) = (9 + 1) + 5$

6. $5 \times (3 \times 8) \times 4 = (5 \times 3) \times (8 \times 4)$ is an example of the

   A. Commutative Property of Multiplication.
   B. Associative Property of Multiplication.
   C. Identity Property of Multiplication.
   D. Distributive Property.

7. $26 - 8 \div 2 =$

   A. 9
   B. 11
   C. 20
   D. 22

8. $4(5-3) \div 4 + 3^2 =$

   A. $\frac{8}{13}$
   B. 8
   C. 11
   D. $13\frac{1}{4}$

9. $8 \times 3 + 9$

   A. 33
   B. 29
   C. 75
   D. 96

10. $3^2 + 4 \times 18 \div 9$

    A. 4
    B. 14
    C. 17
    D. 26

11. $|-5| - |-2| =$

    A. 3
    B. -3
    C. 7
    D. -7

12. $-12 - 8 =$

    A. 20
    B. 4
    C. -4
    D. -20

13. $(-7) + (+10) =$

    A. 17
    B. 3
    C. −3
    D. −17

14. Which of the following is an example of the <u>Associative Property of Multiplication</u>?

    A. $(6 \times 5) \times 9 = 6 \times (5 \times 9)$
    B. $7 + 11 = 11 + 7$
    C. $5(4 + 9) = (5 \times 4) + (5 \times 9)$
    D. $8 + (2 + 3) = (8 + 2) + 3$

15. $4 \times 18 - 9 \div 3^2 =$

    A. 4
    B. 7
    C. 68
    D. 71

16. $-|5| + |-6| =$

    A. 1
    B. −1
    C. 11
    D. −11

17. $14 - 5 + 2 \times 3$

    A. −7
    B. 14
    C. 15
    D. 32

18. $3^2 + 6(5 - 2) =$

    A. 24
    B. 27
    C. 33
    D. 44

19. $7 + 4 \times 5$

    A. 16
    B. 27
    C. 33
    D. 55

20. $6 + 18 \times 4^2 - 4 =$

    A. 146
    B. 188
    C. 290
    D. 380

21. $|-8| - |4| =$

    A. 4
    B. −4
    C. 12
    D. −12

22. Which of the following shows the <u>Identity Property of Addition</u>?

    A. $a + 0 = a$
    B. $a + b = b + a$
    C. $a + (-a) = 0$
    D. $a + (b + c) = (a + b) + c$

23. Which of the following shows the <u>Inverse Property of Multiplication</u>?

    A. $4 \times 1 = 4$
    B. $4 \times (2 \times 1) = (4 \times 2) \times 1$
    C. $4 \times 2 = 2 \times 4$
    D. $4 \times \frac{1}{4} = 1$

24. $-|8| - |-2| =$

    A. 10
    B. −10
    C. 6
    D. −6

25. $24 - 10 \div (3 - 5) + 4^2$

    A. 23
    B. 25
    C. 35
    D. 45

Chapter 13

INTRODUCTION TO ALGEBRA

ALGEBRA VOCABULARY

Vocabulary Word	Example	Definition
variable	$4x$ (x is the variable)	a letter that can be replaced by a number
coefficient	$4x$ (4 is the coefficient)	a number multiplied by a variable or variables
term	$5x^2 + x - 2$ terms	numbers or variables separated by $+$ or $-$ signs
constant	$5x + 2y + 4$ constant	a term that does not have a variable
numerical expression	$2 + 6 - 5$	2 or more terms using only constants (numbers)
algebraic expression	$2x + 5^2 - 7$	2 or more terms that include one or more variables
sentence	$2x = 7$ or $5 \leq x$	two algebraic expressions connected by $=, <, >, \leq, \geq,$ or $\neq$
equation	$4x = 8$	a sentence with an equal sign
inequality	$7x < 30$ or $x \neq 6$	a sentence with one of the following signs: $\neq, <, >, \leq,$ or $\geq$
base	$6^3 \leftarrow$ base	the number used as a factor
exponent	$6^3 \leftarrow$ exponent	the number of times the base is multiplied by itself

SUBSTITUTING NUMBERS FOR VARIABLES

These problems may look difficult at first glance, but they are very easy. Simply replace the variable with the number the variable is equal to and solve.

EXAMPLE 1: In the following problems, substitute 10 for *a*.

	PROBLEM	CALCULATION	SOLUTION
1.	$a + 1$	Simply replace the *a* with 10. $10 + 1$	11
2.	$17 - a$	$17 - 10$	7
3.	$9a$	This means multiply. 9×10	90
4.	$\frac{30}{a}$	This means divide. $30 \div 10$	3
5.	a^3	$10 \times 10 \times 10$	1000
* 6.	$5a + 6$	$(5 \times 10) + 6$	56

* **Note:** Be sure to do all multiplying and dividing before adding and subtracting.

EXAMPLE 2: In the following problems, let $x = 2$, $y = 4$, and $z = 5$.

	PROBLEM	CALCULATION	SOLUTION
1.	$5xy + z$	$5 \times 2 \times 4 + 5$	45
2.	$xz^2 + 5$	$2 \times 5^2 + 5 = 2 \times 25 + 5$	55
3.	$\frac{yz}{x}$	$(4 \times 5) \div 2 = 20 \div 2$	10

In the following problems $t = 7$. Solve the problems.

1. $t + 3 =$ _____
2. $12 - t =$ _____
3. $\frac{21}{t} =$ _____
4. $3t - 5 =$ _____
5. $t^2 + 1 =$ _____
6. $2t - 4 =$ _____
7. $2t \div 3 =$ _____
8. $\frac{t^2}{7} =$ _____
9. $5t + 6 =$ _____

In the following problems $a = 4$, $b = -2$, $c = 5$, and $d = 10$. Solve the problems.

10. $4a + 2c =$ _____
11. $3bc - d =$ _____
12. $\frac{ac}{d} =$ _____
13. $d - 2a =$ _____
14. $c^2 - 4a =$ _____
15. $a^2 - b =$ _____
16. $abd =$ _____
17. $5c - ad =$ _____
18. $cd + bc =$ _____
19. $d - a^2 =$ _____
20. $\frac{6b}{a} =$ _____
21. $9a + b =$ _____
22. $5 + 3bc =$ _____
23. $d^2 + d + 1 =$ _____
24. $7b - 3a =$ _____

UNDERSTANDING ALGEBRA WORD PROBLEMS

The biggest challenge to solving word problems is figuring out whether to add, subtract, multiply, or divide. Below is a list of key words and their meanings. This list does not include every situation you might see, but it includes the most common examples.

Words Indicating Addition	Example	Add
and	6 and 8	6 + 8
increased	The original price of $15 increased by $5	15 + 5
more	3 coins and 8 more	3 + 8
more than	Josh has 10 points. Will has 5 more than Josh.	10 + 5
plus	8 baseballs plus 4 baseballs	8 + 4
sum	the sum of 3 and 5	3 + 5
total	the total of 10, 14, and 15	10 + 14 + 15

Words Indicating Subtraction	Example	Subtract
decreased	$16 decreased by $5	16 − 5
difference	the difference between 18 and 6	18 − 6
less	14 days less 5	14 − 5
less than	Joe completed 2 laps less than Mike's 9.	*9 − 2
left	Ray sold 15 out of 35 tickets. How many did he have left?	*35 − 15
lower than	This month's rain fall is 2 inches lower than last month's rain fall of 3 inches.	*3 − 2
minus	15 minus 6	15 − 6

* In subtraction word problems, you cannot always subtract the numbers in the order that they appear in the problem. Sometimes the first number should be subtracted from the last. You must read each problem carefully.

Word Indicating Multiplication	Example	Multiply
double	Her $1000 profit doubled in a month.	1000 × 2
half	Half of the $600 collected went to charity.	$\frac{1}{2}$ × 600
product	The product of 4 and 8	4 × 8
times	Todd scored 3 times as many points as Ted who only scored 4.	3 × 4
triple	The bacteria tripled its original colony of 10,000 in just one day.	3 × 10,000
twice	Ron has 6 CDs. Tom has twice as many.	2 × 6

Words Indicating Division	Example	Divide
divide into, by, or among	The group of 70 divided into 10 teams.	70 ÷ 10 or $\frac{70}{10}$
quotient	The quotient of 30 and 6	30 ÷ 6 or $\frac{30}{6}$

Match the phrase on the left with the correct algebraic expression on the right. The answers on the right will be used more than once.

1. _____ 2 more than y A. $y - 2$

2. _____ 2 divided into y B. $2y$

3. _____ 2 less than y C. $y + 2$

4. _____ twice y D. $\frac{y}{2}$

5. _____ the quotient of y and 2   E. $2 - y$

6. _____ y increased by 2

7. _____ 2 less y

8. _____ the product of 2 and y

9. _____ y decreased by 2

10. _____ y doubled

11. _____ 2 minus y

12. _____ the total of 2 and y

Now practice writing parts of algebraic expressions from the following word problems.

EXAMPLE: the product of 3 and a number, t Answer: $3t$

13. 3 less than x _____   23. bacteria culture, b, doubled _____

14. y divided among 10 _____   24. triple John's age, y _____

15. the sum of t and 5 _____   25. a number, n, plus 4 _____

16. n minus 14 _____   26. quantity, t, less 6 _____

17. 5 times k _____   27. 10 divided by a number, x _____

18. the total of z and 12 _____ 28. n feet lower than 10 _____

19. double the number b _____   29. 3 more than p _____

20. x increased by 1 _____   30. the product of 4 and m _____

21. the quotient of t and 4 _____ 31. a number, y, decreased by 20 _____

22. half of a number y _____   32. 5 times as much as x _____

If a word problem contains the word "sum" or "difference," put the numbers that "sum" or "difference" refer to in parentheses to be added or subtracted first. Do not separate them. Look at the examples below.

EXAMPLES:	RIGHT	WRONG
sum of 2 and 4, times 5	$5(2+4) = 30$	$2 + 4 \times 5 = 22$
the sum of 4 and 6, divided by 2	$\dfrac{(4+6)}{2} = 5$	$4 + \dfrac{6}{2} = 7$
4 times the difference between 10 and 5	$4(10-5) = 20$	$4 \times 10 - 5 = 35$
20 divided by the difference between 4 and 2	$\dfrac{20}{(4-2)} = 10$	$20 \div 4 - 2 = 3$
the sum of x and 4, multiplied by 2	$2(x+4) = 2x + 8$	$x + 4 \times 2 = x + 8$

Change the following phrases into algebraic expressions.

1. 5 times the sum of x and 6

2. the difference between 5 and 3, divided by 4

3. 30 divided by the sum of 2 and 3

4. twice the sum of 10 and x

5. the difference between x and 9, divided by 10

6. 7 times the difference between x and 4

7. 9 multiplied by the sum of 3 and 4

8. the difference between x and 5, divided by 6

9. x divided by the sum of 4 and 9

10. x minus 5, times 10

11. 100 multiplied by the sum of x and 6

12. twice the difference between 3 and x

13. 4 times the sum of 5 and 1

14. 5 times the difference between 4 and 2

15. 12 divided by the sum of 2 and 4

16. four minus x, multiplied by 2

Look at the examples below for more phrases that may be used in algebra word problems.

EXAMPLES:

one-half of the sum of x and 4	$\frac{1}{2}(x+4)$ or $\frac{x+4}{2}$
six more than four times a number, x	$6 + 4x$
100 decreased by the product of a number, x, and 5	$100 - 5x$
ten less than the product of 3 and x	$3x - 10$

Change the following phrases into algebraic expressions

1. one-third of the sum of x and 5

2. three more than the product of a number, x, and 7

3. ten less than the sum of t and 4

4. the product of 4 and n, minus 3

5. 15 less than the sum of 3 and x

6. the difference of 10, and 3 times a number, n

7. one-fifth of t

8. the product of 3 and x, minus 14

9. x times the difference between 4 and x

10. five plus the quotient of x and 6

11. the sum of 5 and x, divided by 2

12. one less than the product of 3 and x

13. 5 increased by one-half of a number, n

14. 10 more than twice x

15. six subtracted from four times m

16. 8 times x, subtracted from 20

SETTING UP ALGEBRA WORD PROBLEMS

So far, you have seen only the first part of algebra word problems. To complete an algebra problem, an equal sign must be added. The words "**is**" or "**are**" as well as "**equal(s)**" signal that you should add an equal sign.

EXAMPLE: Double Jake's age, x, minus 4 is 22.

$$2x - 4 = 22$$

Translate the following word problems into algebra problems. DO NOT find the solutions to the problems yet.

1. Triple the original number, n, is 2,700.

2. The product of a number, y, and 5 is equal to 15.

3. Four times the difference of a number, x, and 2 is 20.

4. The total, t, divided into 5 groups is 45.

5. The number of parts in inventory, p, minus 54 parts sold today is 320.

6. One-half an amount, x, added to $50 is $262.

7. One hundred seeds divided by 5 rows equals n number of seeds per row.

8. A number, y, less than 50 is 82.

9. His base pay of $200 increased by his commission, x, is $500.

10. Seventeen more than half a number, h, is 35.

11. This month's sales of $2,300 are double January's sales, x.

12. The quotient of a number, y, and 4 is 32.

13. Six less a number, a, is 12.

14. Four times the sum of a number, y, and 10 is 48.

15. We started with x number of students. When 5 moved away, we had 42 left.

16. A number, b, divided into 36 is 12.

MATCHING ALGEBRAIC EXPRESSIONS

Match each set of algebraic expressions with the correct phrase underneath them.

1. ____ $2x + 5$
2. ____ $2(x + 5)$
3. ____ $2x - 5$
4. ____ $2(x - 5)$

A. twice the sum of x and 5
B. five less than the product of 2 and x
C. five more than the product of 2 and x
D. two times the difference of x and 5

---

5. ____ $4(y - 2)$
6. ____ $\dfrac{y - 2}{4}$
7. ____ $4y - 2$
8. ____ $\dfrac{y}{4} - 2$

A. two less than the product of y and 4
B. the difference of y and 2 divided by 4
C. two less than one-fourth of y
D. four times the difference of y and 2

---

9. ____ $5y + 8$
10. ____ $5(y + 8)$
11. ____ $8y + 5$
12. ____ $8(y + 5)$

A. eight times the sum of y and 5
B. eight more than the product of 5 and y
C. five more than eight times y
D. five multiplied by the sum of y and 8

---

13. ____ $9 - x = 7$
14. ____ $x - 9 = 7$
15. ____ $9 - 7 = x$

A. nine less than x is 7
B. nine less x is 7
C. the difference between 9 and 7 is x

---

16. ____ $\dfrac{n + 5}{2} = 10$
17. ____ $\dfrac{n}{2} + 5 = 10$
18. ____ $\tfrac{1}{2}n - 5 = 10$

A. one-half the sum of n and 5 is 10
B. five less than half of n is 10
C. five added to half of n is 10

---

19. ____ $x + \dfrac{4}{5} = 8$
20. ____ $\dfrac{x}{5} + 4 = 8$
21. ____ $\dfrac{x + 4}{5} = 8$

A. the sum of x and 4, divided by 5 is 8
B. x added to the quotient of 4 and 5 is 8
C. four more than x divided by 5 is 8

---

22. ____ $7t + 1 = 5$
23. ____ $7(t + 1) = 5$
24. ____ $7t = 5$

A. one more than seven times t is 5
B. seven times the sum of t and 1 is 5
C. the product of seven and t is 5

CHAPTER REVIEW

Solve the following problems using $x = 2$

1. $3x + 4 =$ ____
2. $\dfrac{6x}{4} =$ ____
3. $x^2 - 5 =$ ____
4. $\dfrac{x^3 + 8}{2} =$ ____
5. $12 - 3x =$ ____
6. $x - 5 =$ ____
7. $-5x + 4 =$ ____
8. $9 - x =$ ____
9. $2x + 2 =$ ____

Solve the following problems. Let $w = -1$, $y = 3$, $z = 5$

10. $5w - y =$ ____
11. $wyz + 2 =$ ____
12. $z - 2w =$ ____
13. $\dfrac{3z + 5}{wz} =$ ____
14. $\dfrac{6w}{y} + \dfrac{z}{w} =$ ____
15. $25 - 2yz =$ ____
16. $-2y + 3 =$ ____
17. $4w - (yw) =$ ____
18. $7y - 5z =$ ____

Write out the algebraic expression or equation given in each word problem.

19. three less the sum of x and 3

20. double Amy's age, a

21. the number of bacteria, b, tripled

22. five less than the product of 5 and y

23. half of a number, n, less 15

24. the quotient of a number, x, and 6

25. one-third the sum of x and 4 is 20

26. ten divided by a number, y, is 2

27. four less a number, n, is 14

28. a number, x, divided into 16 is 4

29. two times a number, y, divided by 3 is 6

30. the difference of x and 5, divided by 2 is 6

CHAPTER TEST

1. Which expression means the same as this phrase?

   8 less than 4 times a number

   A. $8 - x = 4$
   B. $8 + 4x$
   C. $8(x + 4)$
   D. $4x - 8$

2. Which of the following algebraic expressions corresponds to:

   "the product of 5 and x divided by 3 fewer than y"

   A. $\dfrac{5x}{y - 3}$
   B. $\dfrac{5}{x(y - 3)}$
   C. $\dfrac{5x}{3 - y}$
   D. $\dfrac{5x}{y + 3}$

3. If $c = 5$ and $d = 3$, evaluate

   $c^2 - 3d$

   A. 1
   B. 14
   C. 16
   D. 19

4. In the following problem, $a = -2$, $b = 4$ and $c = 5$. Solve the problem.

   $5 + 3a - 4b =$

   A. 14
   B. -17
   C. -49
   D. -180

5. Choose the phrase that means the same as: $5a - 2$

   A. two less than a increased by five
   B. the product of five and a plus two
   C. two less than the difference of 5 and a
   D. two less than the product of 5 and a

6. Choose the statement that means the same as: **the difference of 6 and y, divided by 3 is 12.**

   A. $\dfrac{6 - y}{3} = 12$
   B. $\dfrac{6}{y - 3} = 12$
   C. $\dfrac{y - 6}{3} = 12$
   D. $6 - y = \dfrac{12}{3}$

7. Choose the statement that means the same as: $4(x + 3) = 16$

   A. Four times the sum of three and x is sixteen.
   B. Four more than the sum of three and x is sixteen.
   C. The product of four and x, increased by three is sixteen.
   D. The difference of four and x increased by three is sixteen.

8. Which algebraic expression matches the statement?

   Seventeen is two more than the product of eight and y.

   A. $17 + 2 = 8y$
   B. $17 = 8y + 2$
   C. $17 = y(8 + 2)$
   D. $17 + 2 = 8y$

9. Which algebraic expression matches the following statement?

    Nineteen more than half a number, *m*, is 39.

    A. $\dfrac{19m}{2} = 39$

    B. $19 + m = 39$

    C. $\dfrac{19}{2} + m = 39$

    D. $\dfrac{m}{2} + 19 = 39$

10. Which algebraic expression means the same as the following.

    Take half of *t* and add it to the number 7.

    A. $\dfrac{t+7}{2}$

    B. $\dfrac{t}{2} + 7$

    C. $2t + 7$

    D. $t + \dfrac{7}{2}$

11. Which of the following equations shows that 5 times *x* is 3 more than 2 times *y*?

    A. $5x + 3 = 2y$
    B. $5x = 3 - 2y$
    C. $x = 5(3 + 2y)$
    D. $5x = 3 + 2y$

12. Which value would replace *n* in the following number sentence?

    $4(n - 2) = 16$

    A. 4
    B. 6
    C. 2
    D. 8

13. Choose the phrase that means the same as $3(x + 4)$.

    A. three times a number, *x*, plus 4.
    B. the product of 3 and a number, *x*, added to 4
    C. three times the sum of 4 and a number, *x*.
    D. the product of a number, *x*, and 4 multiplied by 3.

14. In the following problem, $x = 1$, $y = -2$, and $z = 5$. Solve the problem.

    $$\dfrac{yz + 2x}{y}$$

    A. −4
    B. 4
    C. −6
    D. 6

15. Daniel's bonus at work this year is $1000.00 added to 20% of his yearly salary. If *x* represents his yearly salary, which of the following algebraic expressions should be used to figure the amount of his bonus.

    A. $0.2(1000 + x)$

    B. $\dfrac{1000 + x}{0.2}$

    C. $0.2x + 1000$

    D. $\dfrac{x}{0.2} + 1000$

16. Which of the following equations shows that two times the difference of a number, *b*, and 6 is 50?

    A. $2(b + 6) = 50$
    B. $2b - 6 = 50$
    C. $2(b - 6) + 50$
    D. $2(b - 6) = 50$

Chapter 14

ALGEBRA

You have been solving algebra problems since second grade by filling in blanks. For example, 5 + ____ = 8. The answer is 3. You can solve the same kind of problems using algebra. The problems only look a little different because the blank has been replaced with a letter. The letter is called a **variable**.

SOLVING ONE-STEP ALGEBRA PROBLEMS WITH ADDITION AND SUBTRACTION

EXAMPLE 1: Arithmetic   5 + ____ = 14
           Algebra $5 + x = 14$

The goal in any algebra problem is to move all the numbers to one side of the equal sign and have the letter (called a **variable**) on the other side. In this problem, the 5 and the "x" are on the same side. The 5 is added to x. To move it, do the **opposite** of **add**. The **opposite** of **add** is **subtract**, so subtract 5 from both sides of the equation. Now the problem looks like this:

$$\begin{array}{r} 5 + x = 14 \\ -5 -5 \\ \hline x = 9 \end{array}$$

To check your answer, put 9 in the place of x in the original problem. Does 5 + 9 = 14? Yes, it does.

EXAMPLE 2: $y - 16 = 27$ Again, the 16 has to move. To move it to the other side of the equation, we do the **opposite** of **subtract**. We **add** 16 to both sides.

$$\begin{array}{r} y - 16 = 27 \\ +16 +16 \\ \hline y = 43 \end{array}$$

Check by putting 43 in place of the y in the original problem. Does 43 − 16 = 27? Yes.

Solve the problems below.

1. $n + 9 = 27$
2. $12 + y = 55$
3. $51 + v = 67$
4. $f + 16 = 31$
5. $15 + x = 24$
6. $w - 14 = 89$
7. $t - 26 = 20$
8. $m - 12 = 17$
9. $k - 5 = 29$
10. $a + 17 = 45$
11. $d + 26 = 56$
12. $15 + x = 56$
13. $t - 16 = 28$
14. $m + 14 = 37$
15. $y - 21 = 29$
16. $f + 7 = 31$
17. $r - 12 = 37$
18. $h - 17 = 22$
19. $x - 37 = 46$
20. $r - 11 = 28$

MORE ONE-STEP ALGEBRA PROBLEMS

Sometimes the answer to the algebra problem is a negative number. The problems work the same. Study the **examples** below.

EXAMPLES:

$$n + 8 = 6$$
$$-8 \quad -8$$
$$n = (-2)$$

$$x - 10 = -14$$
$$+10 \quad +10$$
$$x = (-4)$$

$$y - (-6) = (-2)$$
$$+(-6) \quad +(-6)$$
$$y = (-8)$$

$$(-4) + m = 9$$
$$+4 \quad +4$$
$$m = 13$$

Solve the problems below.

1. $w + 4 = (-6)$
2. $q - 8 = (-9)$
3. $j + 7 = 1$
4. $y + 14 = 6$
5. $k - 5 = (-8)$
6. $h - 7 = (-2)$
7. $7 + d = (-3)$
8. $(-7) + k = (-4)$

9. $(-4) + h = 8$
10. $(-5) + g = -2$
11. $y - (-8) = (-5)$
12. $x + 1 = (-2)$
13. $w - (-3) = 8$
14. $a - (-6) = 12$
15. $z + 8 = 3$
16. $(-4) + m = 4$

17. $(-10) + r = (-2)$
18. $(-6) + b = 9$
19. $p - (-7) = (-2)$
20. $q - 5 = (-11)$
21. $x + 17 = (-4)$
22. $(-5) + x = 14$
23. $(-17) + r = -12$
24. $(-2) + y = 19$

25. $t - (-2) = (-8)$
26. $d - 9 = (-16)$
27. $x + 16 = -3$
28. $w + 14 = (-2)$
29. $q + 12 = (-5)$
30. $j + (-7) = (-1)$
31. $y - (-9) = 8$
32. $x + 12 = 6$

ONE-STEP ALGEBRA PROBLEMS
WITH MULTIPLICATION AND DIVISION

Solving one-step algebra problems with multiplication and division are just as easy as adding and subtracting. Again, you perform the **opposite** operation. If the problem is a **multiplication** problem, you **divide** to find the answer. If it is a **division** problem, you **multiply** to find the answer. Carefully read the examples below, and you will see how easy they are.

EXAMPLE 1: $4x = 20$ ($4x$ means 4 times x. 4 is the **coefficient** of x.)

The goal is to get the numbers on one side of the equal sign and the variable x on the other side. In this problem, the **4** and x are on the same side of the equal sign. The **4** has to be moved over. $4x$ means 4 times x. The opposite of **multiply** is **divide**. If we divide both sides of the equation by **4**, we will find the answer.

$4x = 20$ We need to divide both sides by 4.

This means divide by 4. → $\frac{4x}{4} = \frac{20}{4}$ We see that $1x = 5$ so $x = 5$

When you put 5 in place of x in the original problem, it is correct. $4 \times 5 = 20$

EXAMPLE 2: $\frac{y}{4} = 2$ This problem means y divided by 4 is equal to 2. In this case, the opposite of **divide** is **multiply**. We need to multiply both sides of the equation by **4**.

$4 \times \frac{y}{4} = 2 \times 4$ so $y = 8$

When you put 8 in place of y in the original problem, it is correct. $\frac{8}{4} = 2$

Solve the problems below.

1. $2x = 14$
2. $\frac{w}{5} = 11$
3. $3h = 45$
4. $10y = 30$

5. $5a = 60$
6. $\frac{x}{3} = 9$
7. $6d = 66$
8. $\frac{w}{9} = 3$

9. $7r = 98$
10. $\frac{y}{3} = 2$
11. $\frac{x}{4} = 35$
12. $\frac{r}{4} = 7$

13. $8t = 96$
14. $\frac{z}{2} = 15$
15. $\frac{n}{9} = 5$
16. $4z = 24$

17. $6d = 84$
18. $\frac{t}{3} = 3$
19. $\frac{m}{6} = 9$
20. $9p = 72$

Sometimes the answer to the algebra problem is a **fraction**. Read the example below, and you will see how easy it is.

> **EXAMPLE**
>
> $4x = 5$ Problems like this are solved just like the problems on the previous page. The only difference is that the answer is a **fraction**.
>
> In this problem, the 4 is **multiplied** by x. To solve, we need to divide both sides of the equation by 4.
>
> $4x = 5$ Now **divide** by 4. $\dfrac{4x}{4} = \dfrac{5}{4}$ Now cancel $\dfrac{\cancel{4}x}{\cancel{4}} = \dfrac{5}{4}$ So $x = \dfrac{5}{4}$
>
> When you put $\dfrac{5}{4}$ in place of x in the original problem, it is correct.
>
> $4 \times \dfrac{5}{4} = 5$ Now cancel. ⟶ $\cancel{4} \times \dfrac{5}{\cancel{4}} = 5$ So $5 = 5$

Solve the problems below. Some of the answers will be fractions. Some answers will be integers.

1. $2x = 3$
2. $4y = 5$
3. $5t = 2$
4. $12j = 144$
5. $9a = 72$
6. $8y = 16$
7. $7x = 21$

8. $4z = 64$
9. $7x = 126$
10. $6p = 10$
11. $2r = 9$
12. $5x = 11$
13. $15m = 180$
14. $5h = 21$

15. $3y = 8$
16. $2t = 10$
17. $3b = 2$
18. $5c = 14$
19. $4d = 3$
20. $5z = 75$
21. $9y = 4$

22. $7d = 12$
23. $2w = 15$
24. $9g = 81$
25. $6a = 12$
26. $2p = 16$
27. $15w = 3$
28. $5x = 13$

200

MULTIPLYING AND DIVIDING WITH NEGATIVE NUMBERS

EXAMPLE 1: $-3x = 15$ In the problem, -3 is **multiplied** by x. To find the solution, we must do the opposite. The opposite of **multiply** is **divide**. We must **divide** both sides of the equation by -3.

$$\frac{-3x}{-3} = \frac{15}{-3}$$ Then cancel. $$\frac{-3x}{-3} = \frac{\overset{5}{\cancel{15}}}{\underset{1}{\cancel{-3}}}$$ $x = -5$

EXAMPLE 2: $\dfrac{y}{-4} = -20$ In this problem, y is **divided** by -4. To find the answer, do the opposite. **Multiply** both sides by -4.

$$-4 \times \frac{y}{-4} = (-20) \times (-4) \quad \text{so} \quad y = 80$$

EXAMPLE 3: $-6a = 2$ The answer to an algebra problem can also be a negative fraction.

$$\frac{-6a}{-6} = \frac{2}{-6} \;\leftarrow\; \text{reduce so} \quad a = \frac{1}{-3} \; \text{or} \; -\frac{1}{3}$$

> **Note:** A negative fraction can be written several different ways.
> $$\frac{1}{-3} = \frac{-1}{3} = -\frac{1}{3} = -\left(\frac{1}{3}\right)$$
> All mean the same thing

Solve the following problems. Reduce any fractions to lowest terms.

1. $2z = -6$
2. $\dfrac{y}{5} = 20$
3. $-16k = 4$
4. $4x = -24$
5. $\dfrac{t}{7} = -4$

6. $\dfrac{x}{-6} = 5$
7. $\dfrac{w}{-11} = 9$
8. $5y = -35$
9. $\dfrac{x}{-4} = -9$
10. $7t = -2$

11. $-8z = 32$
12. $-15w = -60$
13. $\dfrac{y}{-9} = 2$
14. $\dfrac{d}{5} = 7$
15. $-12v = 36$

16. $\dfrac{t}{-2} = -15$
17. $-20x = -4$
18. $-2b = -18$
19. $\dfrac{d}{-7} = 6$
20. $\dfrac{a}{11} = -3$

21. $-4x = -3$
22. $-12y = 7$
23. $\frac{a}{-2} = 22$
24. $-18b = 6$
25. $13a = -39$

26. $\frac{b}{-2} = -14$
27. $-24x = -6$
28. $-6p = 42$
29. $\frac{x}{-23} = -1$
30. $7x = -7$

31. $-9y = -1$
32. $\frac{d}{5} = -10$
33. $\frac{z}{-13} = -2$
34. $-5c = 45$
35. $2d = -3$

36. $-8d = -12$
37. $-24w = 9$
38. $\frac{y}{-5} = -5$
39. $-9a = -18$
40. $\frac{p}{-2} = 15$

VARIABLES WITH A COEFFICIENT OF NEGATIVE ONE

The answer to an algebra problem should not have a negative sign in front of the variable. For example, the problem $-x = 5$ is not completely solved. Study the examples below to learn how to finish solving this problem.

EXAMPLE 1: $-x = 5$ $\quad$ $-x$ means the same thing as $-1x$ or -1 times x. To solve this problem, **multiply** by -1.

$$(-1)(-1x) = (-1)(5) \quad \text{so} \quad x = -5$$

EXAMPLE 2: $-y = -3$ $\quad$ Solve the same way.

$$(-1)(-y) = (-1)(-3) \quad \text{so } y = 3$$

Solve the following problems.

1. $-w = 14$
2. $-c = 20$
3. $-x = -15$
4. $-d = 5$

5. $-x = -25$
6. $-y = -16$
7. $-t = 62$
8. $-b = -30$

9. $-p = -34$
10. $-m = 11$
11. $-w = 17$
12. $-y = 52$

13. $-v = -9$
14. $-k = 13$
15. $-q = 7$
16. $-t = 92$

TWO-STEP ALGEBRA PROBLEMS

In the following two-step algebra problems, **additions** and **subtractions** are performed **first** and *then* division.

EXAMPLE 1: $-4x + 7 = 31$

Step 1: Subtract 7 from both sides.

$$-4x + 7 = 31$$
$$\underline{\; -7 \;\; -7}$$
$$-4x = 24$$

Step 2: Divide both sides by -4.

$$\frac{-4x}{-4} = \frac{24}{-4} \qquad \text{so, } x = -6$$

EXAMPLE 2: $-8 - y = 12$

Step 1: Add 8 to both sides.

$$-8 - y = 12$$
$$\underline{+8 +8}$$
$$-y = 20$$

Step 2: REMEMBER: To finish solving an algebra problem with a negative sign in front of the variable, multiply both sides by -1.

$$(-1)(-y) = (-1)(20) \quad \text{so, } y = -20$$

Solve the two-step algebra problems below.

1. $2x - 2 = -14$
2. $3y - 1 = 32$
3. $11 - 2t = 1$
4. $6p - 3 = -39$
5. $10 - m = -70$
6. $12x + 8 = 20$
7. $3y - 3 = 21$

8. $3x - 11 = 16$
9. $5y - 4 = 49$
10. $7d - 2 = 47$
11. $8h + 8 = -8$
12. $-4b - 15 = -3$
13. $10 - t = 50$
14. $4a + 3 = -5$

15. $-g - 5 = -21$
16. $-3k - 2 = 25$
17. $6 - 10r = 66$
18. $2y - 12 = 12$
19. $3f + 13 = 67$
20. $18 - x = -10$
21. $6y + 2 = 20$

22. $20t + 13 = 33$
23. $12y + 9 = 153$
24. $25p - 18 = 32$
25. $30h + 15 = 75$
26. $-7 + 3w = -61$
27. $5 - y = 14$
28. $2t + 3 = 19$

ALGEBRA WORD PROBLEMS

You saw the following problems in the previous chapter. Translate each word problem into an algebra problem and then solve.

1. Triple the original number, n, is 2,700.

2. The product of a number, y, and 5 is equal to 15.

3. Four times the difference of a number, x, and 2 is 20.

4. The total, t, divided into 5 groups is 9.

5. The number of parts in inventory, p, minus 54 parts sold today is 320.

6. One-half an amount, x, added to $50 is $262.

7. One hundred seeds divided by 5 rows equals n number of seeds per row.

8. A number, y, less than 50 is 82.

9. His base pay of $200 increased by his commission, x, is $500.

10. Seventeen more than half a number, h, is 35.

11. This month's sales of $2,500 are double January's sales, x.

12. The quotient of a number, w, and 4 is 32.

13. Six less a number d, is 12.

14. Four times the sum of a number, y, and 10 is 45.

15. We started with x number of students. When 5 moved away, we had 42 left.

16. A number, b, divided by 6 is 12.

GRAPHING INEQUALITIES

An inequality is a sentence that contains a ≠, <, >, ≤, or ≥ sign. Look at the following graphs of inequalities on a number line.

NUMBER LINE

$x < 3$ is read "x is less than 3".

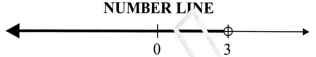

There is no line under the < sign, so the graph uses an **open** endpoint to show x is less than 3 but does not include 3.

$x \leq 5$ is read "x is less than or equal to 5".

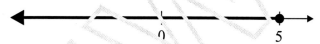

If you see a line under < or > (≤ or ≥), the endpoint is filled in. The graph uses a **closed** circle because the number 5 **is** included in the graph.

$x > -2$ is read "x is greater than −2".

$x \geq 1$ is read "x is greater than or equal to 1".

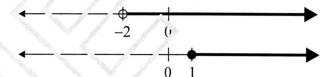

Inequalities can also have more than one part. For example:

$-2 \leq x < 4$ is read "x is greater than or equal to −2 but x is less than 4".

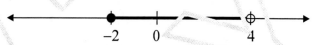

$x < 1$ or $x \geq 4$ is read "x is less than 1, or x is greater than or equal to 4".

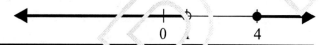

Graph the solution sets of the following inequalities.

1. $x > 8$
2. $x \leq 5$
3. $-5 < x < 1$
4. $x > 7$
5. $1 \leq x < 4$
6. $x < -2$ and $x > 1$
7. $x \geq 10$
8. $x < 4$
9. $x \leq 3$ and $x \geq 5$
10. $x < 1$ and $x > 1$

Give the inequality represented by each of the following number lines.

11. 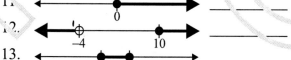 _____
12. _____
13. _____
14. _____
15. _____

16. 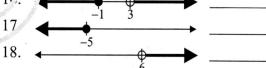 _____
17. _____
18. _____
19. 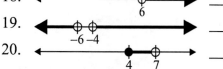 _____
20. _____

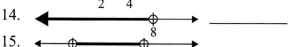

SOLVING INEQUALITIES BY ADDITION AND SUBTRACTION

If you add or subtract the same number to both sides of an inequality, the inequality remains the same. It works just like an equation.

EXAMPLE: Solve and graph the solution set for $x - 2 \leq 5$.

Step 1: Add 2 to both sides of the inequality.

$$\begin{array}{r} x - 2 \leq 5 \\ +2 +2 \\ \hline x \leq 7 \end{array}$$

Step 2: Graph the solution set for the inequality.

Solve and graph the solution set for the following inequalities.

1. $x + 5 > 3$
2. $x - 10 < 5$
3. $x - 2 \leq 1$
4. $9 + x \geq 7$
5. $x - 4 > -2$
6. $x + 11 \leq 20$
7. $x - 3 < -12$
8. $x + 6 \geq -3$
9. $x + 12 \leq 8$
10. $15 + x > 5$
11. $x - 6 < -2$
12. $x + 7 \geq 4$
13. $14 + x \leq 8$
14. $x - 8 > 24$
15. $x + 1 \leq 12$
16. $11 + x \geq 11$
17. $x - 3 < 17$
18. $x + 9 > -4$
19. $x + 6 \leq 11$
20. $x - 8 \geq 19$

SOLVING INEQUALITIES BY MULTIPLICATION AND DIVISION

If you multiply or divide both sides of an inequality by a **positive** number, the inequality symbol stays the same. However, if you multiply or divide both sides of an inequality by a **negative** number, **you must reverse the direction of the inequality symbol**.

EXAMPLE 1: Solve and graph the solution set for $4x \leq 20$.

Step 1: Divide both sides of the inequality by 4. $\quad \dfrac{\overset{1}{\cancel{4}}x}{\underset{1}{\cancel{4}}} \leq \dfrac{\overset{5}{\cancel{20}}}{\cancel{4}}$

Step 2: Graph the solution. $\quad x \leq 5$

EXAMPLE 2: Solve and graph the solution set for $6 > -\dfrac{x}{3}$.

Step 1: Multiply both sides of the inequality by -3 and **reverse** the direction of the symbol.

$$(-3) \times 6 < -\dfrac{x}{3} \times (-3)$$

Step 2: Graph the solution. $-18 < x$

Solve and graph the following inequalities.

1. $\dfrac{x}{5} > 4$

2. $2x \leq 24$

3. $-6x \geq 36$

4. $\dfrac{x}{10} > -2$

5. $-\dfrac{x}{4} > 8$

6. $-7x \leq -49$

7. $-3x > 18$

8. $-\dfrac{x}{7} \geq 9$

9. $9x \leq 54$

10. $\dfrac{x}{8} > 1$

11. $-\dfrac{x}{9} \leq 3$

12. $-4x < -12$

13. $-\dfrac{x}{2} \geq -20$

14. $10x \leq 30$

15. $\dfrac{x}{12} > -4$

16. $-6x < 24$

CHAPTER REVIEW

Solve the following algebra equations.

1. $9y = 27$
2. $2x + 5 = 35$
3. $2 - 5t = 22$
4. $\dfrac{c}{-12} = -1$

5. $-9k = -54$
6. $9 + h = -4$
7. $5p = 2$
8. $p - 14 = 1$

9. $20t - 4 = -24$
10. $z - (-1) = -12$
11. $2w - 23 = 27$
12. $-4 + f = -12$

13. $-1 + 5g = 34$
14. $\dfrac{m}{7} = -3$
15. $d + (-2) = 24$
16. $7 - p = 47$

Solve and graph the following inequalities.

17. $d - 11 > 14$
18. $\dfrac{a}{3} \geq -12$
19. $-6b > 24$
20. $-10d \leq -40$

21. $\dfrac{w}{3} < -6$
22. $n - 21 \geq -7$
23. $-a \leq 26$
24. $5y \leq -45$

Solve the following algebra word problems.

25. Four less than a number, x, is 5.

26. The number of cracked eggs, x, is five less than the total of 25.

27. One-half the student body, n, is 250.

28. Five times the difference of 3 and x is 15.

29. Ten divided into the sum of distance, t, and 15 miles is 25 miles.

30. Five times the number, x, is 25.

CHAPTER TEST

1. If $x + 37 = 78$, what is x?

    A. -36
    B. 36
    C. -41
    D. 41

2. If $x - 12 = 42$, what is x?

    A. 64
    B. 30
    C. 54
    D. 34

3. Which value of x, will make the sentence true?

    $$4x - 2 = 22$$

    A. 5
    B. 6
    C. 16
    D. 20

4. If $4x + 6 = 54$, what is x?

    A. 7
    B. 12
    C. 13
    D. 8

5. If $\frac{n}{6} = -3$, what is n?

    A. -48
    B. 14
    C. 8
    D. -2

6. If $12x - 8 = 124$, what is x?

    A. 8
    B. 10
    C. 14
    D. 11

7. Solve for x: $x - 7 = (-9)$

    A. 2
    B. -2
    C. -11
    D. -16

8. What would replace n in this number sentence to make the sentence true?

    $$2n - 8 = 16$$

    A. 4
    B. 6
    C. 8
    D. 12

9. Dane is three years older than twice his brother's age, b. Which algebraic expression below represents Dane's age?

    A. $b + 3$
    B. $2b + 3$
    C. $3 - 2b$
    D. $2b - 3$

10. The sum of the distance, n, and 24 miles, divided by 7 is 11. Find n.

    A. 18
    B. 31
    C. 53
    D. 77

11. Find v: $v + 3 \geq 24$

    A. $v \leq 21$
    B. $v \leq 27$
    C. $v \geq 21$
    D. $v \geq 27$

12. Find x: $x - 4 < 9$

    A. $x > 5$
    B. $x < 5$
    C. $x > 13$
    D. $x < 13$

13. Find y: $-4y < 56$

    A. $y > -14$
    B. $y > 14$
    C. $y < -14$
    D. $y < 14$

14. Find t: $\dfrac{t}{2} \leq -8$

    A. $t \leq -4$
    B. $t \geq -4$
    C. $t \leq -16$
    D. $t \geq -16$

15. Find c: $\dfrac{c}{-2} > -6$

    A. $c > -12$
    B. $c < -12$
    C. $c > 12$
    D. $c < 12$

16. [number line with closed circle at 6, shaded from 0 to 6]

    Which inequality is represented on the number line above?

    A. $x > 6$
    B. $x \geq 6$
    C. $x < 6$
    D. $x \leq 6$

17. [number line with closed circle at 5 shaded left, open circle at 9 shaded right]

    Which solution set is graphed on the number line above?

    A. $x \leq 9$ and $x > 5$
    B. $x < 9$ and $x > 5$
    C. $x \geq 9$ and $x < 5$
    D. $x > 9$ and $x \leq 5$

18. [number line with open circle at -3, closed circle at 4, shaded between]

    Which inequality is represented on the number line above?

    A. $-3 > x \geq 4$
    B. $-3 \leq x < 4$
    C. $-3 < x \leq 4$
    D. $-3 \leq x \leq 4$

19. Which number line below represents the following inequality?

    $$-2 < x \leq 3$$

    A. [open circle at -2, closed at 3, shaded between]
    B. [closed circle at -2, open at 3, shaded between]
    C. [open circle at -2, closed at 3, shaded between]
    D. [closed circle at -2, open at 3, shaded between]

20. Which number line below represents the following inequality?

    $$x \geq 5$$

    A. [shaded left from 5]
    B. [shaded left to open circle at 5]
    C. [closed circle at 5, shaded right]
    D. [open circle at 5, shaded right]

21. Which number line below represents the following inequality?

    $$x \leq 16 \text{ and } x > 9$$

    A. [open circle at 9, closed at 16, shaded between]
    B. [closed circle at 9, open at 16, shaded between]
    C. [open circle at 9, closed at 16, shaded between]
    D. [closed circle at 9, open at 16, shaded between]

22. Which number line below represent the following inequality?

    $$x < -4 \text{ and } x \geq 5$$

    A. [open circle at -4, closed at 5, shaded between]
    B. [closed circle at -4, open at 5, shaded between]
    C. [open circle at -4, closed at 5, shaded right]
    D. [closed circle at -4, open at 5]

Chapter 15: FUNCTIONS, NUMBER PATTERNS, PERMUTATIONS, AND COMBINATIONS

FUNCTION MACHINES

A **function** is a rule given as an equation, table, or graph that shows the relationship between a number in one set to a unique number in another set. A function machine illustrates this relationship. In a function machine, for each number put into the machine, only one unique number is output from the machine.

Function Machine

7 → input
−5 → Function → 2 output

A 7 is put into the machine. The function of the machine is to subtract 5, so the output is 2.
$7 - 5 = 2$

For each function machine below, give the output.

1. 10 input → + 6 function → _____ output

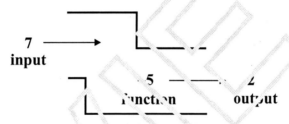

2. 4 → × 7 → _____

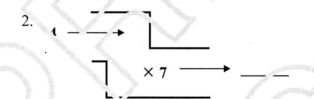

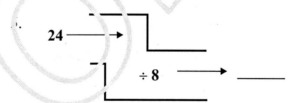

3. −5 → + 2 → _____

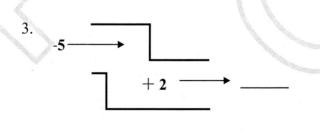

6. −9 → × 5 → _____

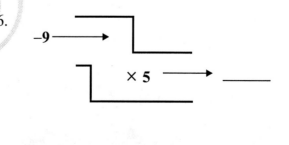

MORE FUNCTION MACHINES

A function machine can also look like this:

```
   8   | Input
 x + 4 | Function
  12   | Output
```

For these function machines, replace x with the input, 8, and solve. $8 + 4 = 12$

For each function machine below, find the output.

1. Input: 14, Function: $x - 3$, Output: ___

2. Input: -1, Function: $x + 5$, Output: ___

3. Input: 5, Function: $6x$, Output: ___

4. Input: -7, Function: $5 - x$, Output: ___

5. Input: 12, Function: $\frac{x}{3}$, Output: ___

6. Input: -9, Function: $10x$, Output: ___

7. Input: 5, Function: $10 \div x$, Output: ___

8. Input: -3, Function: $2x - 3$, Output: ___

9. Input: 3, Function: $-1 - x$, Output: ___

10. Input: 6, Function: $9x$, Output: ___

11. Input: 14, Function: $x \div 2$, Output: ___

12. Input: 25, Function: $30 - x$, Output: ___

13. Input: 6, Function: $x - 7$, Output: ___

14. Input: -3, Function: $-2x$, Output: ___

15. Input: 24, Function: $\frac{x}{8}$, Output: ___

212 Copyright © American Book Company

ALGEBRAIC FUNCTIONS

In algebra, if there is only one unique value of y for each value of x in an equation, the equation is a **function**.

EXAMPLE:

Rule $y = 4x$	
x	y
0	0
1	4
2	8
3	12
4	16

The chart at the left shows that for each value of x, there is only one value for y. When you replace x with 0, $y = 0$. When you replace x with 1, $y = 4$, and so on.

Find the unique value of y for each value of x for each of the functions below.

1.

Rule $y = 3x - 2$	
x	y
0	
1	
2	
3	
4	

4.

Rule $y = 5x + 2$	
x	y
1	
3	
5	
7	
9	

7.

Rule $y = 5x$	
x	y
0	
1	
2	
3	
4	

2.

Rule $y = 2(x - 4)$	
x	y
2	
3	
4	
5	
6	

5.

Rule $y = 3(5 + x)$	
x	y
−10	
−5	
0	
5	
10	

8.

Rule $y = 10 - 2x$	
x	y
−8	
−1	
0	
4	
8	

3.

Rule $y = \dfrac{2(x-4)}{2}$	
x	y
0	
2	
6	
8	
10	

6.

Rule $y = \dfrac{3(x-2)}{5}$	
x	y
−8	
−3	
7	
12	
17	

9.

Rule $y = \dfrac{(4+x)}{3}$	
x	y
−7	
−1	
2	
5	
8	

FUNCTION TABLES

Functions can also use a variable such as *n* to be the input of the function and *f(n)*, read "*f* of *n*" to represent the output of the function.

EXAMPLE: Function rule: $3n + 4$

n	f(n)
−1	1
0	4
1	7
2	10

Fill in the tables for each function rule below.

1. rule: $2(n-5)$

n	f(n)
0	
1	
2	
3	

2. rule: $3x(x-4)$

x	f(x)
0	
1	
2	
3	

3. rule: $\dfrac{2-n}{2}$

n	f(n)
0	
2	
4	
6	
8	

4. rule: $2x(x-1)$

x	f(x)
1	
2	
3	
4	

5. rule: $\dfrac{1}{n+3}$

n	f(n)
1	
2	
3	
4	

6. $4x - x$

x	f(x)
−2	
−1	
0	
1	
2	

7. rule: $n(n+2)$

n	f(n)
1	
2	
3	
4	

8. rule: $2x - 3$

x	f(x)
1	
2	
3	
4	

9. rule: $3 - 2n$

n	f(n)
−2	
−1	
0	
1	
2	

IDENTIFYING THE OPERATION

In the equations below, the operation is missing. The numbers for the problem and the answer are given; however, you need to insert a +, −, ×, or ÷ sign to make the equation true.

EXAMPLE 1: $3 \bigcirc 4 = 12$ To make this equation true, fill in a × sign in the circle.

$3 \otimes 4 = 12$

EXAMPLE 2: $18 \bigcirc 9 = 2$ To make this equation true, fill in a ÷ sign in the circle.

$18 \div 9 = 2$

In the problems below, fill in the missing sign of operation.

1. $32 \bigcirc 5 = 27$
2. $15 \bigcirc -3 = -5$
3. $13 \bigcirc 1 = 12$
4. $-6 \bigcirc -8 = 48$
5. $12 \bigcirc 7 = 5$
6. $-12 \bigcirc 4 = -3$
7. $14 \bigcirc 2 = 28$
8. $56 \bigcirc -8 = -7$
9. $-12 \bigcirc -3 = 36$
10. $3 \bigcirc (-4) = 7$
11. $14 \bigcirc 2 = 7$
12. $12 \bigcirc 6 = 6$
13. $12 \bigcirc 7 = 19$
14. $17 \bigcirc 8 = 9$

15. $24 \bigcirc \frac{1}{2} = 12$
16. $15 \bigcirc \frac{1}{3} = 45$
17. $27 \bigcirc 5 = 22$
18. $3(2 \bigcirc 1) = 9$
19. $-9 \bigcirc 5 = -4$
20. $23 \bigcirc 8 = 31$
21. $16 \bigcirc \frac{1}{8} = 2$
22. $\frac{1}{4} \bigcirc \frac{1}{2} = \frac{1}{8}$
23. $\frac{2}{3} \bigcirc \frac{1}{3} = 1$
24. $\frac{1}{2} \bigcirc \frac{1}{4} = \frac{1}{4}$
25. $30 \bigcirc \frac{1}{3} = 10$
26. $2(5 \bigcirc 4) = 2$
27. $\frac{1}{3} \bigcirc \frac{1}{2} = \frac{2}{3}$
28. $2 \bigcirc 5 - 3 = 7$

29. $15 \bigcirc 5 = 20$
30. $2^2 \bigcirc 3^2 = 13$
31. $63 \bigcirc 7^2 = 14$
32. $35 \bigcirc -5 = -7$
33. $8^2 \bigcirc 20 = 44$
34. $4 \bigcirc 5^2 = 29$
35. $54 \div (5 \bigcirc 4) = 6$
36. $5(6 \bigcirc 2) = 20$
37. $(50 \bigcirc 5) \div 2 = 5$
38. $40 \bigcirc 2^3 = 5$
39. $20 \div (10 \bigcirc 5) = 4$
40. $\frac{1}{8} \bigcirc \frac{1}{16} = 2$
41. $25 \bigcirc \frac{1}{5} = 5$
42. $\frac{1}{6} \bigcirc \frac{1}{3} = \frac{1}{2}$

215

NUMBER PATTERNS

In each of the examples below, there is a sequence of numbers that follows a pattern. Think of the sequence of numbers like the output for a function. You must find the pattern (or function) that holds true for each number in the sequence. Once you determine the pattern, you can find the next number in the sequence or any number in the sequence.

	Sequence	Pattern	Next Number	20th number in the sequence
EXAMPLE 1:	3, 4, 5, 6, 7	$n + 2$	8	22

In number patterns, the sequence is the output. The input can be the set of whole numbers starting with 1. But, you must determine the "rule" or pattern. Look at the table below.

input	sequence
1	→ 3
2	→ 4
3	→ 5
4	→ 6
5	→ 7

What pattern or "rule" can you come up with that gives you the first number in the sequence, 3, when you input 1? $n + 2$ will work because when $n = 1$, the first number in the sequence = 3. Does this pattern hold true for the rest of the numbers in the sequence? Yes, it does. When $n = 2$, the second number in the sequence = 4. When $n = 3$, the third number in the sequence = 5, and so on. Therefore, $n + 2$ is the pattern. Even without knowing the algebraic form of the pattern, you could figure out that 8 is the next number in the sequence. To find the 20th number in the pattern, use $n = 20$ to get 22.

	Sequence	Pattern	Next Number	20th number in the sequence
EXAMPLE 2:	1, 4, 9, 16, 25	n^2	36	400
EXAMPLE 3:	−2, −4, −6, −8, −10	$−2n$	−64	−40

Find the pattern and the next number in each of the sequences below.

	Sequence	Pattern	Next Number	20th number in the sequence
1.	−2, −1, 0, 1, 2			
2.	5, 6, 7, 8, 9			
3.	3, 7, 11, 15, 19			
4.	−3, −6, −9, −12, −15			
5.	3, 5, 7, 9, 11			
6.	2, 4, 8, 16, 32			1,048,576
7.	1, 8, 27, 64, 125			
8.	0, −1, −2, −3, −4			
9.	2, 5, 10, 17, 26			
10.	4, 6, 8, 10, 12			

216

OTHER PATTERN PROBLEMS

EXAMPLE 1: In a board game, players move their game pieces clockwise around the corners of a square game board. If the first turn lands on the corner marked W, on which corner will the game piece land after 50 moves?

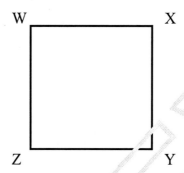

You could keep counting around the square until you get to 50 to see which corner you land on, or you can follow the simple steps below.

Step 1: Determine the pattern. What kind of pattern do you see? The first move is at W, the second move is at X, the third at Y, and the fourth at Z. Then you start over with move 5 = W, 6 = X, 7 = Y, 8 = Z and so on.

Which moves land on W? 1, 5, 9, 13, 17, 21 and so on. What's the pattern?

Moves that land on W follow the pattern $1 + 4 + 4 + 4 + 4 + ...$

Which moves land on X? 2, 6, 10, 14, 18, 22 and so on. What's the pattern?

Moves that land on X follow the pattern $2 + 4 + 4 + 4 + 4 + ...$

Which moves land on Y? 3, 7, 11, 15, 19, 23 and so on. What's the pattern?

Moves that land on Y follow the pattern $3 + 4 + 4 + 4 + 4 + ...$

Which moves land on Z? 4, 8, 12, 16, 20, 24 and so on. What's the pattern?

Moves that land on Z follow the pattern $4 + 4 + 4 + 4 + 4 + ...$

Step 2: In this type of problem, 4 repeats itself in every pattern. To determine which letter the game piece will be on after 50 moves, figure out how many times 4 will go into 50 evenly and what is the remainder.

$50 \div 4 = 12$ with a remainder of 2

So, which letter follows the pattern of 12 4's plus 2? The answer is the letter X. After 50 moves, the game piece will be on letter X.

To think about this problem another way, every four moves will land on the letter Z. After 12 moves of 4 (or 48 moves) the game piece will be on Z. After two more moves (a total of 50 moves), it will be on X.

Find the answer for each of the following pattern problems.

A family plays a board game around a hexagonal board. On each turn, a player moves his game piece clockwise to the next corner of the hexagon.

Answer the following questions using the diagram below. The first question is worked for you.

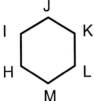

1. The first move lands on corner J. After 75 moves, on which corner will the game piece be?

   J's pattern: $1 + 6 + 6 + ...$
   K's pattern: $2 + 6 + 6 + ...$
   (L's pattern: $3 + 6 + 6 + ...$)
   M's pattern: $4 + 6 + 6 + ...$
   H's pattern: $5 + 6 + 6 + ...$
   I's pattern: $6 + 6 + 6 + ...$

   $75 \div 6 = 12$ remainder 3 L

2. The first move lands on corner M. On the 100th move, on which corner will the game piece land?

3. The first move lands on corner H. On which corner will the game piece be after 240 moves?

4. The first move lands on corner L. Which corner will the game piece land on after 62 moves?

On the school playground, a teacher invents a game played around a circle. Her seven students stand in a circle and pass a ball clockwise to the next student in the circle.

Use the diagram to answer the following questions.

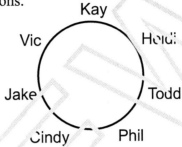

5. If the first pass goes from Kay to Heidi, who will catch the ball on the 65th pass? (Careful: Who is the first to catch the ball?)

6. If the ball starts at Jake and the first pass goes to Vic, who will receive the 120th pass?

7. If the ball starts at Jake, who will throw the ball on the 421st pass?

8. If the ball starts at Cindy, who will receive the ball on the 50th pass? (Careful: Cindy *throws* the first pass. She does not *receive* the first pass.)

PERMUTATIONS

A **permutation** is an arrangement of items in a specific order. If a problem asks how many ways can you arrange 6 books on a bookshelf, it is asking you how many permutations there are for 6 items.

EXAMPLE 1: Ron has 4 items: a model airplane, a trophy, an autographed football, and a toy sports car. How many ways can he arrange the 4 items on a shelf?

Solution: Ron has 4 choices for the first item on the shelf. He then has 3 choices left for the second item. After choosing the second item, he has 2 choices left for the third item and only one choice for the last item. The diagram below shows the permutations for arranging the 4 items on a shelf if he chose to put the trophy first.

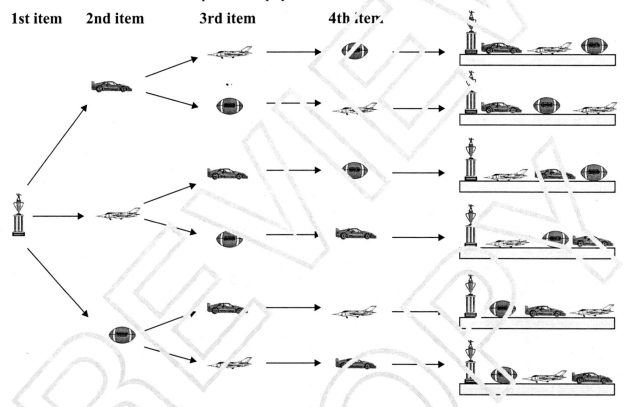

Count the number of permutations if Ron chooses the trophy as the first item. There are 6 permutations. Next, you could construct a pyramid of permutations choosing the model car first. That pyramid would also have 6 permutations. Then, you could construct a pyramid choosing the airplane first. Finally, you could construct a pyramid choosing the football first. You would then have a total of 4 pyramids each having 6 permutations. The total number of permutations is $6 \times 4 = 24$. There are 24 ways to arrange the 4 items on a bookshelf.

You probably don't want to draw pyramids for every permutation problem. What if you want to know the permutations for arranging 30 objects? Fortunately, mathematicians have come up with a formula for calculating permutations.

For the above problem, Ron has 4 items to arrange. Therefore, multiply $4 \times 3 \times 2 \times 1 = 24$. Another way of expressing this calculation is $4!$, stated as 4 factorial. $4! = 4 \times 3 \times 2 \times 1$.

EXAMPLE 2: How many ways can you line up 6 students?

Solution: The number of permutations for 6 students = 6! = 6 × 5 × 4 × 3 × 2 × 1 = 720. There are 6 choices for the first position, 5 for the second, 4 for the third, 3 for the fourth, and 2 for the fifth, and 1 for the sixth.

Work the following permutation problems.

1. How many ways can you arrange 5 books on a bookshelf?

2. Myra has 6 novels to arrange on a book shelf. How many ways can she arrange the novels?

3. Seven sprinters signed up for the 100 meter dash. How many ways can the seven sprinters line up on the start line?

4. Keri wants an ice-cream cone with one scoop of chocolate, one scoop of vanilla, and one scoop of strawberry. How many ways can the scoops be arranged on the cone?

5. How many ways can you arrange the letters A, B, C, and D?

6. At Sam's party, the DJ has four song requests. In how many different orders can he play the 4 songs?

7. Yvette has 5 comic books. How many different ways can she stack the comic books?

8. Sandra's couch can hold three people. How many ways can she and her two friends sit on the couch?

9. How many ways can you arrange the numbers 2, 3, 5?

10. At a busy family restaurant, four tables open up at the same time. How many different ways can the hostess seat the next four families waited to be seated?

MORE PERMUTATIONS

EXAMPLE: If there are 6 students, how many ways can you line up any 4 of them?

Solution: Multiply $6 \times 5 \times 4 \times 3 = 360$. There are 6 choices for the first position in line, 5 for the second position, 4 for the third position, and 3 for the last position. There are 360 ways to line up 4 of the 6 students.

Find the number of permutations for each of the problems below.

1. How many ways can you arrange 4 out of 8 books on a shelf?

2. How many 3 digit numbers can be made using the numbers 2, 3, 5, 8, and 9?

3. How many ways can you line up 4 students out of a class of 20?

4. Kim worked in the linen department of a store. Eight new colors of towels came in. Her job was to line up the new towels on a long shelf. How many ways can she arrange the 8 colors?

5. Terry's CD player holds 5 CD's. Terry owns 12 CD's. How many different ways can he arrange his CD's in the CD player?

6. Erik has 11 shirts he wears to school. How many ways can he choose a different shirt to wear on Monday, Tuesday, Wednesday, Thursday, and Friday?

7. Deb has a box of 12 markers. The art teacher told her to choose three markers and line them up on her desk. How many ways can she line up 3 markers from the 12?

8. Jeff went into an ice cream store serving 32 flavors of ice cream. He wanted a cone with two different flavors. How many ways could he order 2 scoops of ice cream, one on top of the other?

9. In how many ways can you arrange any 3 letters from the 26 letters in the alphabet?

COMBINATIONS

In a **permutation**, objects are arranged in a particular order. In a **combination**, the order does not matter. In a **permutation**, if someone picked two letters of the alphabet, **k, m** and **m, k** would be considered 2 different permutations. In a **combination, k, m** and **m, k** would be the same combination. A different order does not make a new combination.

EXAMPLE: How many combinations of 3 letters from the set {a, b, c, d, e} are there?

Step 1: Find the **permutation** of 3 out of 5 objects.

Step 2: Divide by the permutation of the **number of objects** to be chosen from the total (3). This step eliminates the duplicates in finding the permutations.

$$\frac{5 \times 4 \times 3}{3 \times 2 \times 1} = 10$$

Step 3: Cancel common factors and simplify.

Find the number of combinations for each problem below.

1. How many combinations of 4 numbers can be made from the set of numbers {2, 4, 6, 7, 8, 9}?

2. Johnston Middle School wants to choose 3 students at random from the 7th grade to take an opinion poll. There are 124 seventh graders in the school. How many different groups of 3 students could be chosen? (Use a calculator for this one.)

3. How many combinations of 3 students can be made from a class of 20?

4. Fashion Ware catalog has a sweater that comes in 8 colors. How many combinations of 2 different colors does a shopper have to choose from?

5. Angelo's Pizza offers 10 different pizza toppings. How many different combinations can be made of pizzas with four toppings?

6. How many different combinations of 5 flavors of jelly beans can you make from a store that sells 25 different flavors of jelly beans?

7. The track team is running the relay race in a competition this Saturday. There are 4 members of the track team. The relay race requires 4 runners. How many combinations of 4 runners can be formed from the track team?

8. Kerri got to pick 2 prizes from a grab bag containing 12 prizes. How many combinations of 2 prizes are possible?

MORE COMBINATIONS

Another kind of combination involves selection from several categories.

EXAMPLE: At Joes's Deli, you can choose from 4 kinds of bread, 5 meats, and 3 cheeses when you order a sandwich. How many different sandwiches can be made with Joe's choices for breads, meats, and cheeses if you choose 1 kind of bread, 1 meat, and 1 cheese for each sandwich?

JOE'S SANDWICHES

Breads	**Meats**	**Cheeses**
White	Roast Beef	Swiss
Pumpernickel	Corned Beef	American
Light rye	Pastrami	Mozzarella
Whole wheat	Roast Chicken	
	Roast Turkey	

Solution: Multiply the number of choices in each category. There are 4 breads, 5 meats, and 3 cheeses, so $4 \times 5 \times 3 = 60$. There are 60 combinations of sandwiches.

Find the number of combinations that can be made in each of the problems below.

1. Angie has 4 pairs of shorts, 6 shirts, and 2 pairs of tennis shoes. How many different outfit combinations can be made with Angie's clothes?

   _____

2. Raymond has 7 baseball caps, 2 jackets, 10 pairs of jeans, and 2 pairs of sneakers. How many combinations of the 4 items can he make?

   _____

3. Claire has 6 kinds of lipstick, 4 eye shadows, 2 kinds of lip liner, and 2 mascaras. How many combinations can she use to make up her face?

   _____

4. Clarence's dad is ordering a new truck. He has a choice of 5 exterior colors, 3 interior colors, 2 kinds of seats, and 3 sound systems. How many combinations does he have to pick from?

   _____

5. A fast food restaurant has 8 kinds of sandwiches, 3 kinds of french fries, and 5 kinds of soft drinks. How many combinations of meals could you order if you ordered a sandwich, fries, and a drink?

   _____

6. In summer camp, Tyrone can choose from 4 outdoor activities, 3 indoor activities, and 3 water sports. He has to choose one of each. How many combinations of activities can he choose?

   _____

7. Jackie won a contest at school and gets to choose one pencil and one pen from the school store and an ice-cream from the lunch room. There are 5 colors of pencils, 3 colors of pens, and 4 kinds of ice-cream. How many combinations of prize packages can she choose?

   _____

CHAPTER REVIEW

Give the output for each of the following function machines.

1. 6 → [− 2] → ____

2. 9 → [x + 9] → ____

3. 14 → [+ 5] → ____

4. 10 → [x − 6] → ____

5. 15 → [+ 3] → ____

6. 22 → [x − 1] → ____

Fill in the following function tables.

7. Rule $y = 3(x - 5)$

x	y
1	
2	
3	
4	
5	

8. Rule $y = \dfrac{(6 - x)}{x}$

x	y
−2	
−1	
1	
2	

9. Rule $y = x(x + 1)$

x	y
1	
2	
3	
4	
5	

10. Rule: $\dfrac{(4 - 2n)}{2}$

n	f(n)
0	
1	
2	
3	
4	

11. Rule: $2n(n + 1)$

n	f(n)
0	
1	
2	
3	
4	

12. Rule: $6n - 3$

n	f(n)
2	
3	
4	
5	
6	

Fill in the missing sign of operation.

13. 5 ◯ 9 = 45

14. 24 ◯ 6 = 4

15. $\frac{1}{3}$ ◯ 9 = 3

16. 10 ◯ 4 = 14

17. $\frac{1}{8}$ ◯ $\frac{3}{8}$ = $\frac{1}{2}$

18. 32 ◯ 2 = 16

19. −3 ◯ 8 = 5

20. $\frac{1}{7}$ ◯ $\frac{1}{2}$ = $\frac{2}{7}$

21. 15 ◯ 5 = 10

Find the pattern for the following number sequences. Then find the requested number in the sequence.

22. 0, 1, 2, 3, 4   pattern _____

23. 0, 1, 2, 3, 4   20th number _____

24. 1, 3, 5, 7, 9   pattern _____

25. 1, 3, 5, 7, 9   25th number _____

26. 3, 6, 9, 12, 15   pattern _____

27. 3, 6, 9, 12, 15   30th number _____

Answer the following pattern problems.

28. Will, Kyle, Stan, and Ric form a square and pass a football to each other in a clockwise direction.

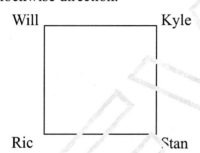

If the first pass goes from Will to Kyle, who will receive the 50th pass?

_____

29. A board game is played around a triangle. Players move their game pieces clockwise around the corners of the triangle.

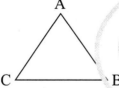

If the first move lands on the corner marked A, on which corner will the 100th move land?

_____

Answer the following permutation and combination problems.

30. Daniel has 7 trophies he has won playing soccer. How many different ways can he arrange them in a row on his bookshelf?

_____

31. Missy has 12 colors of nail polish. She wears 1 color each day, 7 different colors a week. How many combinations of 7 colors can she make before she has to repeat the same 7 colors in a week?

_____

32. Eileen has a collection of 12 antique hats. She plans to donate 5 of the hats to a museum. How many combinations of hats are possible for her donation?

_____

33. Julia has 5 porcelain dolls. How many ways can she arrange 3 of the dolls on a display shelf?

_____

34. Ms. Randai has 10 students. Every day she randomly draws the names of 2 students out of a bag to turn in their homework for a test grade. How many combinations of 2 students can she draw?

_____

35. In the lunch line, students can choose 1 out of 3 meats, 1 out of 4 vegetables, and 1 out of 3 desserts, and 1 out of 5 drinks. How many lunch combinations are there?

_____

CHAPTER TEST

1. What is the output of the following function?

   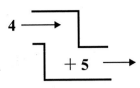

   A. −1
   B. 1
   C. 9
   D. −9

2. What is the output of the following function?

   14 → −1 − x →

   A. 13
   B. −13
   C. 15
   D. −15

3. What is the output of the following function?

   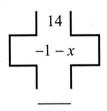

   3 → 5x + 2 →

   A. 7
   B. 10
   C. 15
   D. 17

4. The function rule is $3x(x+5)$

x	f(x)
−2	

   A. −18
   B. 18
   C. −42
   D. 42

5.
Rule $y = \frac{2x+1}{3}$	
x	y
4	

   For the given value of x, what is the value of y?

   A. 3
   B. 4
   C. 8
   D. 9

6. Which operation should replace the ◯ in the following equation?

   $\frac{1}{5}$ ◯ 20 = 4

   A. +
   B. −
   C. ×
   D. ÷

7. Which operation should replace the ◯ in the following equation?

   30 ◯ 60 = $\frac{1}{2}$

   A. +
   B. −
   C. ×
   D. ÷

8. Which of the following could be a pattern for the sequence 6, 10, 18, 34, ... ?

   A. $n + 4$
   B. $4n$
   C. $n^2 + 2$
   D. $2^n + 2$

9. What is the 12th number in the following sequence?

   3, 6, 9, 12, 15, ...

   A. 12
   B. 36
   C. 45
   D. 48

10. What is the 20th number in the following sequence?

    −3, −2, −1, 0, 1, ...

    A. 20
    B. 17
    C. 16
    D. 15

11. What number comes next in the sequence?

    1, 4, 9, 16, 25, ...

    A. 27
    B. 29
    C. 30
    D. 36

12.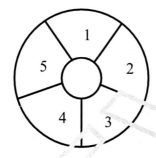

    The first grade class drew the diagram above on the playground. Scott jumped first on number 1. Then he jumped on each numbered section going clockwise around the circle. When he jumped the 27th time, on which numbered section did he land?

    A. 1
    B. 2
    C. 3
    D. 4

13. Andrea has 7 teddy bears in a row on a shelf in her room. How many ways can she arrange the bears in a row on her shelf?

    A. 7
    B. 49
    C. 343
    D. 5,040

14. Adrianna has 4 hats, 8 shirts, and 9 pairs of pants. Choosing one of each, how many different clothes combinations can she make?

    A. 17
    B. 32
    C. 96
    D. 288

15. Andy has 5 favorite songs on one CD. His CD player will allow him to program the songs in any order. How many different ways can he program those 5 songs one time each?

    A. 5
    B. 25
    C. 120
    D. 125

16. Zandra went into a candy store to buy jelly beans. She wanted to choose 3 flavors from the 12 flavors they sold. How many combinations of 3 flavors could she make?

    A. 12
    B. 36
    C. 220
    D. 1,320

17. The buffet line offers 5 kinds of meat, 3 different salads, a choice of 4 desserts, and 5 different drinks. If you chose one food from each category, from how many combinations would you have to choose?

    A. 17
    B. 20
    C. 60
    D. 300

18. How many different ways can Mrs. Smith choose 2 students to go to the library if she has 20 students in her class?

    A. 40
    B. 190
    C. 380
    D. 3,800

Chapter 16

ANGLES

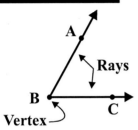

Angles are made up of two rays with a common endpoint. Rays are named by the endpoint, B, and another point on the ray. Ray $\overrightarrow{BA}$ and ray $\overrightarrow{BC}$ share a common endpoint.

Angles are usually named by three capital letters. The middle letter names the vertex. The angle to the left can be named ∠ABC or ∠CBA. An angle can also be named by a lower case letter between the sides, ∠x, or by the vertex alone, ∠B.

A protractor, is used to measure angles. The protractor is divided evenly into a half circle of 180 degrees (180°). When the middle of the bottom of the protractor is placed on the vertex, and one of the rays of the angle is lined up with 0°, the other ray of the angle crosses the protractor at the measure of the angle. The angle below has the ray pointing left lined up with 0° (the outside numbers), and the other ray of the angle crosses the protractor at 55°. The angle measures 55°.

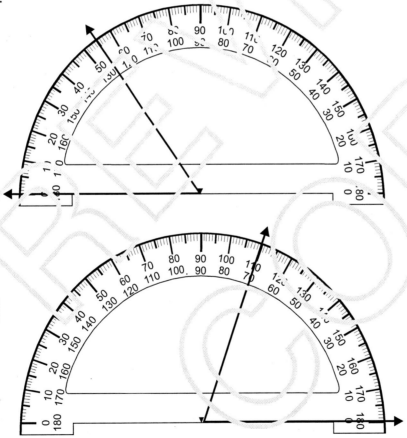

The angle above has the ray pointing right lined up with 0° using the inside numbers. The other ray crosses the protractor and measures the angle at 70°.

TYPES OF ANGLES

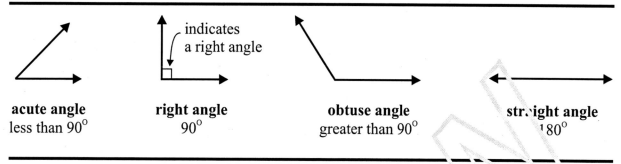

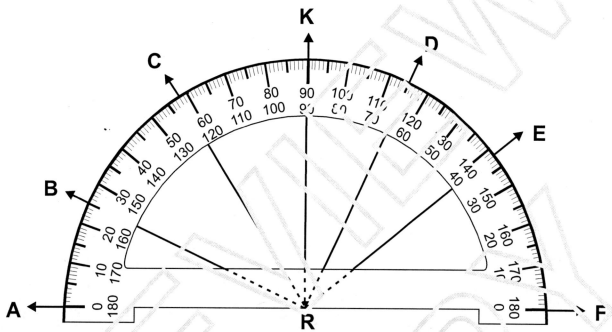

Using the protractor above, find the measure of the following angles. Then tell what type of angle it is: acute, right, obtuse, or straight.

		Measure	Type of Angle
1.	What is the measure of angle ARF?		
2.	What is the measure of angle CRF?		
3.	What is the measure of angle BRF?		
4.	What is the measure of angle ERF?		
5.	What is the measure of angle ARB?		
6.	What is the measure of angle KRA?		
7.	What is the measure of angle CRA?		
8.	What is the measure of angle DRF?		
9.	What is the measure of angle ARD?		
10.	What is the measure of angle FRK?		

MEASURING ANGLES

Estimate the measure of the following angles. Then use your protractor to record the actual measure.

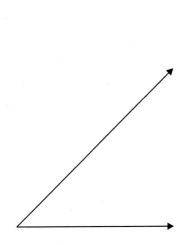

1. Estimate = _____°
   Measure = _____°

4. Estimate = _____°
   Measure = _____°

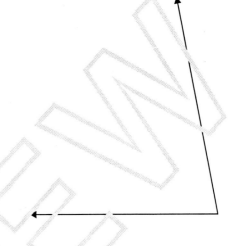

7. Estimate = _____°
   Measure = _____°

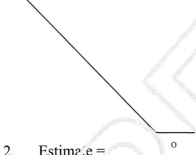

2. Estimate = _____°
   Measure = _____°

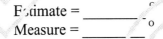

5. Estimate = _____°
   Measure = _____°

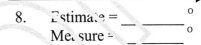

8. Estimate = _____°
   Measure = _____°

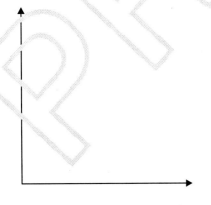

3. Estimate = _____°
   Measure = _____°

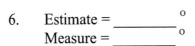

6. Estimate = _____°
   Measure = _____°

9. Estimate = _____°
   Measure = _____°

ADJACENT ANGLES

Adjacent angles are two angles that have the same vertex and share one ray. They do not share space inside the angles.

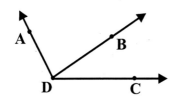

In this diagram, ∠ADB is **adjacent** to ∠BDC.

However, ∠ADB is **not adjacent** to ∠ADC because adjacent angles do not share any space inside the angle.

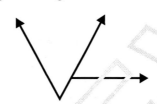

These two angles are **not adjacent**. They share a common ray but do not share the same vertex.

For each diagram below, name the angle that is adjacent to it.

1.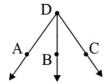

   ∠CDB is adjacent to ∠_____

2.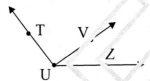

   ∠TUV is adjacent to ∠_____

3.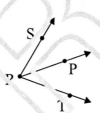

   ∠SRP is adjacent to ∠_____

4.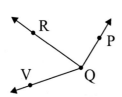

   ∠PQR is adjacent to ∠_____

5.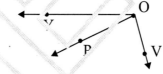

   ∠YOP is adjacent to ∠_____

6.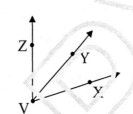

   ∠XVY is adjacent to ∠_____

7.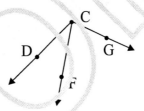

   ∠DCF is adjacent to ∠_____

8.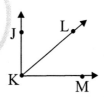

   ∠JKL is adjacent to ∠_____

VERTICAL ANGLES

When two lines intersect, two pairs of vertical angles are formed. Vertical angles are not adjacent. Vertical angles have the same measure.

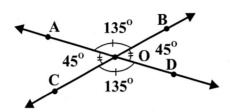

∠AOB and ∠COD are vertical angles. ∠AOC and ∠BOD are vertical angles. **Vertical angles are congruent.** Congruent means they have the same measure.

In the diagrams below, name the second angle in each pair of vertical angles.

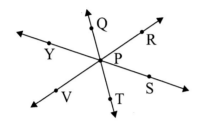

1. ∠YPV _____   4. ∠VPT _____   7. ∠MLN _____   10. ∠GLM _____

2. ∠QPR _____   5. ∠RPT _____   8. ∠KLH _____   11. ∠KLM _____

3. ∠SPT _____   6. ∠YPS _____   9. ∠GLN _____   12. ∠HLG _____

Use the information given to find the measure of each unknown vertical angle.

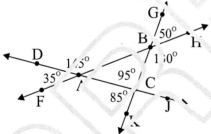 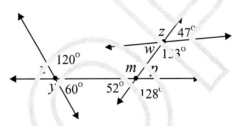

13. ∠CAF = _____   19. ∠x = _____

14. ∠ABC = _____   20. ∠y = _____

15. ∠KCJ = _____   21. ∠z = _____

16. ∠ABG = _____   22. ∠w = _____

17. ∠BCJ = _____   23. ∠m = _____

18. ∠CAB = _____   24. ∠p = _____

232

COMPLEMENTARY AND SUPPLEMENTARY ANGLES

Two angles are **complementary** if the sum of the measures of the angles is 90°.
Two angles are **supplementary** if the sum of the measures of the angles is 180°.

The angles may be adjacent but do not need to be.

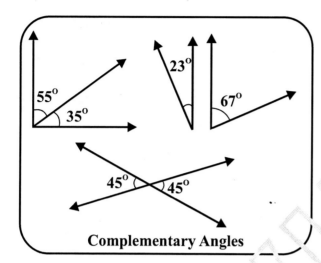
Complementary Angles

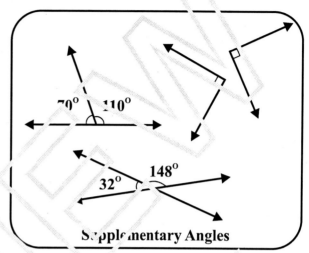
Supplementary Angles

Calculate the measure of each unknown angle.

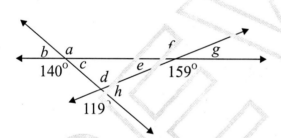

1. ∠a = _____
2. ∠b = _____
3. ∠c = _____
4. ∠d = _____

5. ∠e = _____
6. ∠f = _____
7. ∠g = _____
8. ∠h = _____

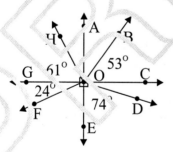

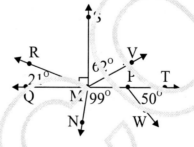

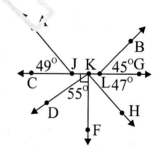

9. ∠AOB = _____
10. ∠COD = _____
11. ∠EOF = _____
12. ∠AOH = _____

13. ∠RMS = _____
14. ∠VMT = _____
15. ∠QMN = _____
16. ∠WPQ = _____

17. ∠AJK = _____
18. ∠CKD = _____
19. ∠FKH = _____
20. ∠BLC = _____

Copyright © American Book Company

CHAPTER REVIEW

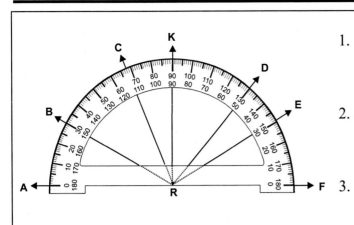

1. What is the measure of ∠DRA?

   _____

2. What is the measure of ∠CRF?

   _____

3. What is the measure of ∠ARB?

   _____

4. Which angle is a supplementary angle to ∠EDF?

   _____

5. What is the measure of angle ∠GDF?

   _____

6. Which 2 angles are right angles?

   _____ and _____

7. What is the measure of ∠EDF?

   _____

8. Which angle is adjacent to ∠BAD?

   _____

9. Which angle is a complementary angle to ∠HAD?

   _____

10. What is the measure of ∠HAB?

    _____

11. What is the measure of ∠CAD?

    _____

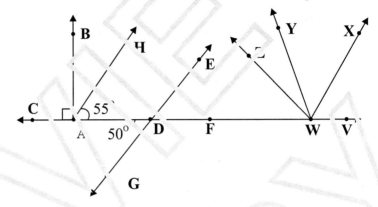

12. What kind of angle is ∠FDA?

    _____

13. What kind of angle is ∠GDA?

    _____

14. Which angles are adjacent to ∠EDA?

    _____ and _____

15. Measure ∠VWX with a protractor.

    _____

16. Measure ∠FWY with a protractor.

    _____

17. Measure ∠VWY with a protractor.

    _____

234 Copyright © American Book Company

CHAPTER TEST

Use the protractor below to answer questions 1 - 3.

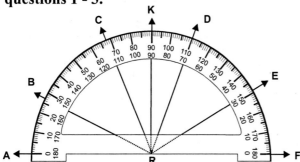

1. What is the measure of ∠ BRF?

   A. 27°
   B. 33°
   C. 153°
   D. 167°

2. What is the measure of ∠ ARC?

   A. 68°
   B. 72°
   C. 112°
   D. 128°

3. What is the measure of ∠ ERF?

   A. 31°
   B. 49°
   C. 149°
   D. 151°

4. Use a protractor to measure the angle below.

   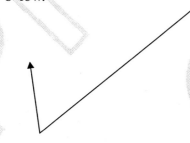

   A. 45°
   B. 60°
   C. 100°
   D. 120°

5. Which of the following is an acute angle?

   A.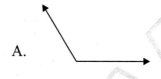

   B.

   C.

   D.

6. Which of the following is a right angle?

   A.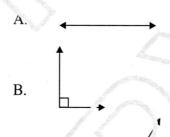

   B.

   C.

   D.

7. Use a protractor to measure the angle below.

   A. 10°
   B. 110°
   C. 170°
   D. 180°

Use the diagram below to answer questions 8 - 9.

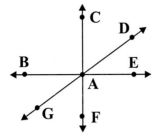

8. Name the angle that is vertical to ∠ EAF.

   A. ∠ EAD
   B. ∠ CAB
   C. ∠ BAG
   D. ∠ GAF

9. Which of the two angles are adjacent?

   A. ∠ GAB and ∠ DAE
   B. ∠ CAG and ∠ CAD
   C. ∠ EAF and ∠ CAB
   D. ∠ CAE and ∠ BAG

Use the diagram below to answer questions 10 - 11.

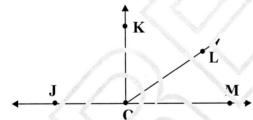

10. Which two angles are supplementary?

    A. ∠ JOL and ∠ KOM
    B. ∠ LOM and ∠ KOL
    C. ∠ JOK and ∠ KOL
    D. ∠ KOM and ∠ JOK

11. Which two angles are complementary?

    A. ∠ JOK and ∠ LOM
    B. ∠ KOM and ∠ LOM
    C. ∠ KOL and ∠ LOM
    D. ∠ JOL and ∠ LOM

Use the diagram below to answer questions 12 - 16.

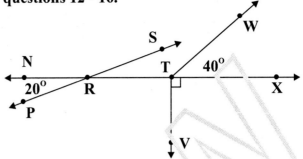

12. What is the measure of ∠ WTN?

    A. 40°
    B. 50°
    C. 140°
    D. 180°

13. What is the measure of ∠ SRT?

    A. 20°
    B. 40°
    C. 90°
    D. 160°

14. What is the measure of ∠ TRP?

    A. 20°
    B. 40°
    C. 90°
    D. 160°

15. Which kind of angle is ∠ XTV?

    A. Right
    B. Obtuse
    C. Acute
    D. Left

16. Which angle is a straight angle?

    A. ∠ NTW
    B. ∠ WTX
    C. ∠ XTN
    D. ∠ PRT

Chapter 17

PLANE GEOMETRY

The following terms are important for understanding the concepts that are presented in this chapter. Most of you have already been introduced to these terms. Rather than defining them in words, they are presented here by example as a refresher.

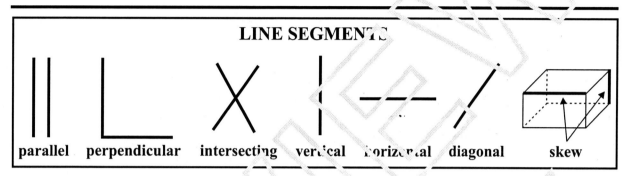

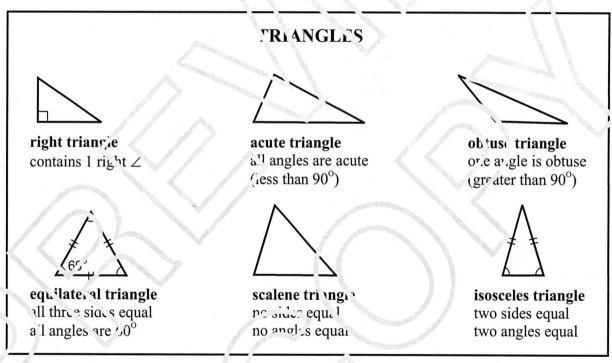

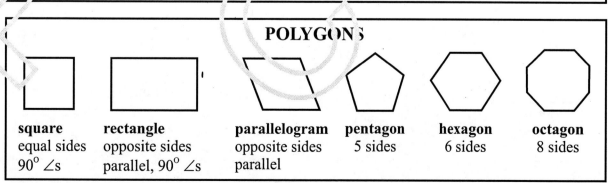

PERIMETER

The **perimeter** is the distance around a polygon. To find the perimeter, add the lengths of the sides.

EXAMPLES:

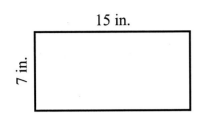

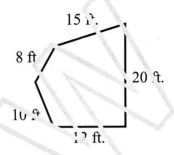

P = 7 + 15 + 7 + 15
P = 44 in.

P = 4 + 6 + 5
P = 15 cm

P = 8 + 15 + 20 + 12 + 10
P = 65 ft.

Find the perimeter of the following polygons.

1. Rectangle: 8 in., 5 in.

2. Pentagon: 3 ft., 2 ft., 2 ft., 5 ft., 5 ft.

3. 32 cm, 29 cm, 29 cm, 33 cm, 10 cm, 35 cm

4. Triangle: 13 cm, 15 cm, 10 cm

5. Trapezoid: 12 in., 8 in., 8 in., 16 in.

6. 9 ft., 7 ft., 5 ft., 6 ft.

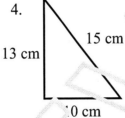

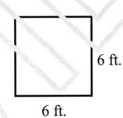

7. Square: 6 ft., 6 ft.

8. 25 cm, 22 cm, 10 cm, 13 cm, 20 cm

9. Octagon: 4 in. (each side)

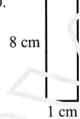

10. Rectangle: 8 cm, 1 cm

11. Triangle: 7 ft., 6 ft., 8 ft.

12. Trapezoid: 7 cm, 28 cm, 30 cm, 24 cm

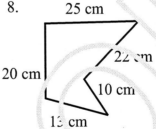

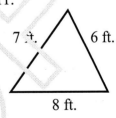

AREA OF SQUARES AND RECTANGLES

Area - area is always expressed in square units such as in^2, cm^2, ft^2, and m^2.

The area, (A), of squares and rectangles equals length (ℓ) times width (w). $A = \ell w$

EXAMPLE:

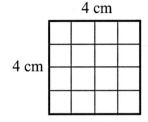

$A = \ell w$
$A = 4 \times 4$
$A = 16 \text{ cm}^2$

If a square has an area of 16 cm^2, it means that it will take 16 squares that are 1 cm on each side to cover the area of a square that is 4 cm on each side.

Find the area of the following squares and rectangles using the formula $A = \ell w$.

1.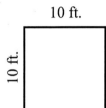

2.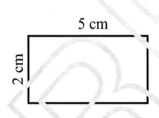

3.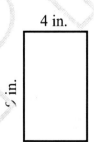

4.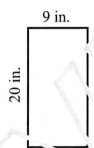

5.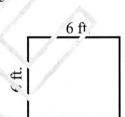

6.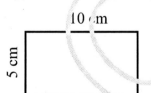

7.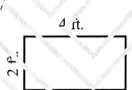

8.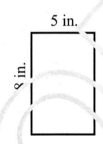

9.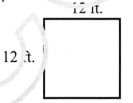

10. 7 cm / 12 cm

11.

12.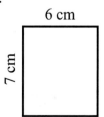

Copyright © American Book Company

AREA OF TRIANGLES

EXAMPLE: Find the area of the following triangle.

The formula for the area of a triangle is:

$A = \frac{1}{2} \times b \times h$

A = area
b = base
h = height

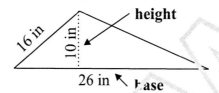

Step 1 Insert measurements from the triangle into the formula. $A = \frac{1}{2} \times 26 \times 10$

Step 2 Cancel and multiply. $A = \frac{1}{\cancel{2}} \times \frac{\cancel{26}^{13}}{1} \times \frac{10}{1} = 130$ in^2

Note: Area is always expressed in square units such as in^2, ft^2, cm^2, or m^2.

Find the area of the following triangles. Remember to include units.

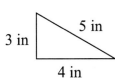
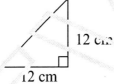
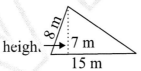
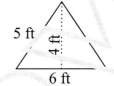

1. _____ in^2 4. _____   7. _____   10. _____

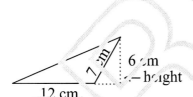
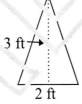
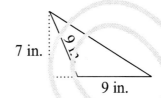
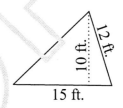

2. _____   5. _____   8. _____   11. _____

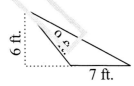
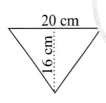
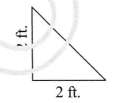
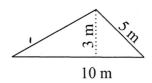

3. _____   6. _____   9. _____   12. _____

PARTS OF A CIRCLE

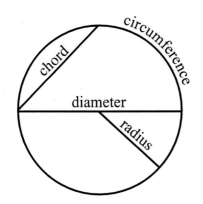 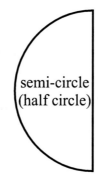 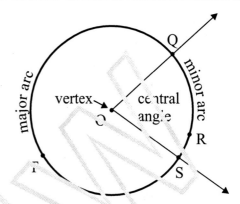

A **central angle** of a circle has the center of the circle as its vertex. The rays of a central angle each contain a radius of the circle. ∠QOS is a central angle.

The points Q and S separate the circle into **arcs**. The arc lies on the circle itself. It does not include any points inside or outside the circle. $\overset{\frown}{QRS}$ or $\overset{\frown}{QS}$ is a **minor arc** because it is less than a semicircle. A minor arc can be named by 2 or 3 points. $\overset{\frown}{QTS}$ is a **major arc** because it is more than a semicircle. A major arc must be named by 3 points.

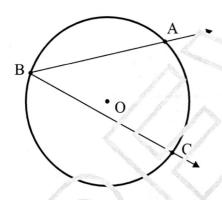

An **inscribed angle** is an angle whose vertex lies on the circle and whose sides contain **chords** of the circle. ∠ABC is an inscribed angle.

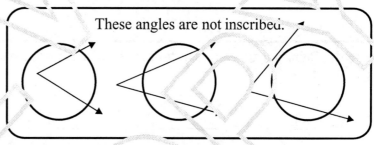

These angles are not inscribed.

Refer to the figure on the right, and answer the following questions.

1. Identify the 2 line segments that are chords of the circle but not diameters. _____ _____

2. Identify the largest major arc of the circle that contains point S. _____

3. Identify the vertex of the circle. _____

4. Identify the inscribed angle. _____

5. Identify the central angle. _____

6. Identify the line segment that is a diameter of the circle. _____

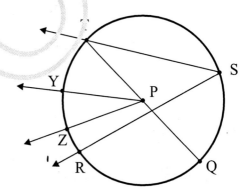

CIRCUMFERENCE

Circumference, C - the distance around the outside of a circle
Diameter, d - a line segment passing through the center of a circle from one side to the other
Radius, r - a line segment from the center of a circle to the edge of the circle
Pi, π - the ratio of the circumference of a circle to its diameter $\pi = 3.14$ or $\pi = \frac{22}{7}$

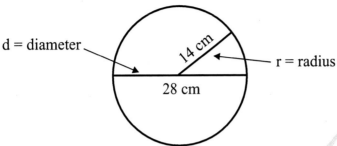

The formula for the circumference of a circle is **C = 2πr** or **C = πd**. (The formulas are equal because the diameter is equal to twice the radius, d = 2r.)

EXAMPLE:	**EXAMPLE:**
Find the circumference of the circle above.	Find the circumference of the circle above.
C = πd Use = 3.14 C = 3.14 × 28 C = 87.92 cm	C = 2πr C = 2 × 3.14 × 14 C = 87.92 cm

---

Use the formulas given above to find the circumference of the following circles. Use π = 3.14.

1. 8 in.
2. 14 ft.
3. 2 cm
4. 6 in.
5. 8 ft.

C = _____   C = _____   C = _____   C = _____   C = _____

Use the formulas given above to find the circumference of the following circles. Use π = $\frac{22}{7}$.

6. 3 ft.
7. 12 in.
8. 6 m
9. 5 cm
10. 16 in.

C = _____   C = _____   C = _____   C = _____   C = _____

AREA OF A CIRCLE

The formula for **area of a circle is** $A = \pi r^2$. The area is how many square units of measure would fit inside a circle.

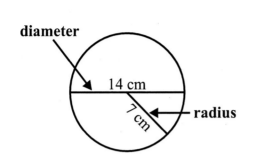

$\pi = \dfrac{22}{7}$ or $\pi = 3.14$

EXAMPLE: Find the area of the circle using both values for π.

Let $\pi = \dfrac{22}{7}$

$A = \pi r^2$

$A = \dfrac{22}{7} \times 7^2$

$A = \dfrac{22}{7} \times \dfrac{49}{1} = 154 \text{ cm}^2$

Let $\pi = 3.14$

$A = \pi r^2$

$A = 3.14 \times 7^2$

$A = 3.14 \times 49 = 153.86 \text{ cm}^2$

Find the AREA of the following circles.

$\pi = 3.14$ $\pi = \dfrac{22}{7}$

1.
   5 in.
   A = _____   A = _____

2.
   10 ft.
   A = _____   A = _____

3.
   8 cm
   A = _____   A = _____

4.
   3 m
   A = _____   A = _____

Fill in the chart below.

	Radius	Diameter	Area $\pi = 3.14$	Area $\pi = \dfrac{22}{7}$
5.	9 ft.			
6.		4 in.		
7.	8 cm			
8.		20 ft.		
9.	14 m			
10.		18 cm		
11.	12 ft.			
12.		6 in.		

AREA AND CIRCUMFERENCE MIXED PRACTICE

1. Find the area of a circle with a diameter of 12 cm. Use $\pi = 3.14$.

2. Which circle has a larger circumference, a circle with a radius of 9 feet or a circle with a diameter of 5 feet?

3. Find the circumference of a circle that is 14 cm in diameter.

    Use $\pi = \frac{22}{7}$.

4. Paula has a round garden. The diameter is 14 feet. What is the area of the garden? Use $\pi = 3.14$

Find the area and circumference of each circle. Use $\pi = 3.14$

5.    22 feet

6.    30 inches

7. 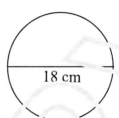   18 cm

A = _____   A = _____   A = _____

C = _____   C = _____   C = _____

Find the area and circumference of each circle. Use $\pi = \frac{22}{7}$

8. 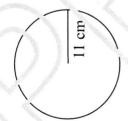   11 cm

9. 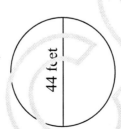   44 feet

10. 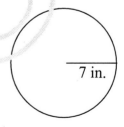   7 in.

A = _____   A = _____   A = _____

C = _____   C = _____   C = _____

TWO-STEP AREA PROBLEMS

Solving the problems below will require two steps. You will need to find the area of two figures, and then either add or subtract the two areas to find the answer. **Carefully read the EXAMPLES below.**

EXAMPLE

Find the area of the living room below.

Figure 1

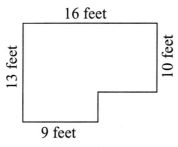

Step 1 Complete the rectangle as in Figure 2, and figure the area as if it were a complete rectangle.

Figure 2

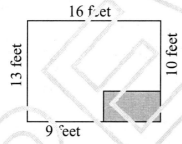

A = length × width
A = 16 × 13
A = 208 ft²

Step 2 Figure the area of the shaded part.

7 × 3 = 21 ft²

Step 3 Subtract the area of the shaded part from the area of the complete rectangle.

208 − 21 = 187 ft²

EXAMPLE

Find the area of the shaded sidewalk.

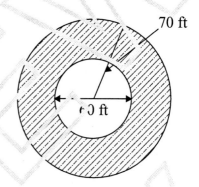

Step 1 Find the area of the outside circle.
π = 3.14
A = 3.14 × 70 × 70
A = 15,386 ft²

Step 2 Find the area of the inside circle.
π = 3.14
A = 3.14 × 30 × 30
A = 2,826 ft²

Step 3 Subtract the area of the inside circle from the area of the outside circle.

15,386 − 2,826 = 12,560 ft²

Find the area of the following figures.

1.

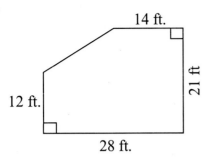

_____

2.

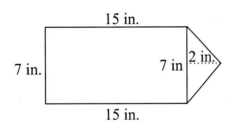

_____

3. What is the area of the shaded circle? Use π = 3.14. Round the answer to the nearest whole number.

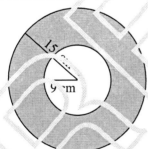

_____

4.

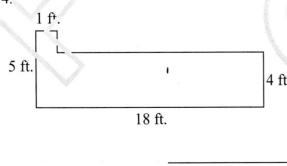

_____

5. What is the area of the rectangle that is shaded? Use π = 3.14 and round to the nearest whole number.

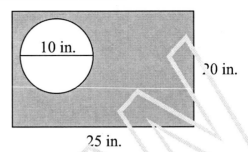

_____

6. What is the area of the shaded part?

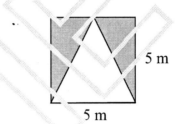

_____

7. What is the area of the shaded part?

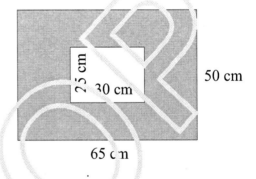

_____

8.

24 m
6 m
12 m
12 m

_____

MEASURING TO FIND PERIMETER AND AREA

Using a ruler, measure the perimeter of the following figures and calculate the area. Be careful to measure in the correct unit. To calculate the area of some of the figures, you will have to use the two-step approach.

1.

    P = _____ cm

    A = _____ cm^2

4.

    P = _____ in

    A = _____ in^2

7.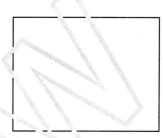

    P = _____ cm

    A = _____ cm^2

2.

    P = _____ in

    A = _____ in^2

5.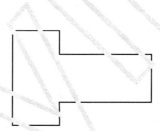

    P = _____ in

    A = _____ in^2

8.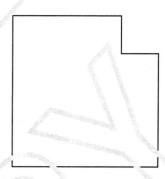

    P = _____ cm

    A = _____ cm^2

3.

    P = _____ in

    A = _____ in^2

6.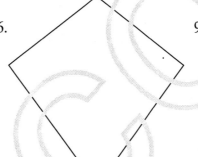

    P = _____ cm

    A = _____ cm^2

9.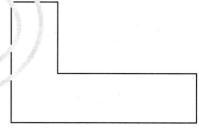

    P = _____ in

    A = _____ in^2

ESTIMATING AREA

Measure the following objects with your ruler and estimate the area to the nearest whole number.

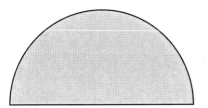

1. Area is about _____ cm²

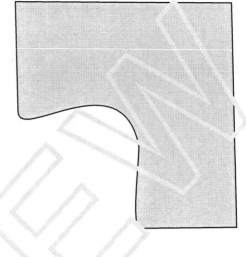

4. Area is about _____ cm²

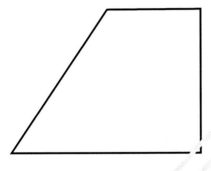

2. Area is about _____ in²

5. Area is about _____ cm².

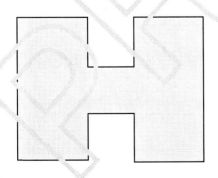

3. Area is about _____ in²

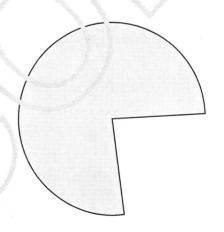

6. Area is about _____ in².

SIMILAR AND CONGRUENT

Similar figures have the same shape but are two different sizes. Their corresponding sides are proportional. **Congruent figures** are exactly alike in size and shape and their corresponding sides are equal. See the examples below.

SIMILAR **CONGRUENT**

Directions: Label each pair of figures below as either **S** if they are similar, **C** if they are congruent, or **N** if they are neither.

1.

2.

3.

4.

5.

6.

7.

8.

9.

10.

11.

12.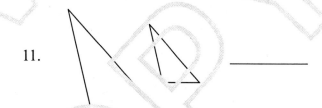

13.

14.

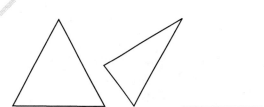

SOLVING PROPORTIONS

Two **ratios (fractions)** that are **equal** to each other are called **proportions.**
For example $\frac{1}{4} = \frac{2}{8}$.

Read the following **Example** to see how to find a number missing from a proportion.

Example: $\frac{5}{15} = \frac{8}{x}$

Step 1 To find x, you first multiply the two numbers that are diagonal to each other. $15 \times 8 = 120$

Step 2 Then divide the product (120) by the other number in the proportion (5). $120 \div 5 = 24$

Therefore $\frac{5}{15} = \frac{8}{x}$ $x = 24$

Practice finding the number missing from the following proportions. First, multiply the two numbers that are diagonal from each other. Then divide by the other number.

1. $\frac{1}{2} = \frac{3}{x}$
2. $\frac{21}{7} = \frac{x}{5}$
3. $\frac{x}{9} = \frac{6}{18}$
4. $\frac{2}{x} = \frac{9}{12}$
5. $\frac{5}{x} = \frac{3}{9}$
6. $\frac{x}{x} = \frac{7}{28}$
7. $\frac{50}{25} = \frac{x}{9}$

8. $\frac{1}{x} = \frac{9}{31}$
9. $\frac{6}{30} = \frac{x}{5}$
10. $\frac{14}{2} = \frac{x}{3}$
11. $\frac{15}{6} = \frac{5}{x}$
12. $\frac{2}{x} = \frac{5}{10}$
13. $\frac{x}{7} = \frac{4}{14}$
14. $\frac{9}{4} = \frac{x}{8}$

15. $\frac{15}{3} = \frac{20}{x}$
16. $\frac{x}{4} = \frac{12}{3}$
17. $\frac{15}{4} = \frac{x}{2}$
18. $\frac{x}{10} = \frac{3}{5}$
19. $\frac{2}{7} = \frac{x}{14}$
20. $\frac{2}{5} = \frac{4}{x}$
21. $\frac{x}{20} = \frac{3}{10}$

SIMILAR TRIANGLES

Two triangles are similar if the measurements of the three angles in both triangles are the same. If the three angles are the same, then their corresponding sides are proportional.

CORRESPONDING SIDES - The triangles below are similar. Therefore, the two shortest sides from each triangle, **c** and **f**, are corresponding. The two longest sides from each triangle, **a** and **d**, are corresponding. And the two medium length sides, **b** and **e**, are corresponding.

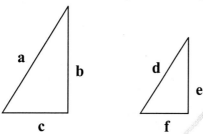

PROPORTIONAL - The corresponding sides of similar triangles are proportional to each other. This means if we know all the measurements of one triangle and only know one measurement of the other triangle, we can figure out the measurements of the other two sides with proportion problems. The two triangles below are similar.

Note: **To set up the proportion correctly, it is important to keep the measurements of each triangle on opposite sides of the equal sign.**

To find the short side:	To find the medium length side:
Step 1 Set up the proportion.	**Step 1** Set up the proportion.
$\dfrac{long\ side}{short\ side}\qquad \dfrac{12}{6} = \dfrac{16}{?}$	$\dfrac{long\ side}{medium}\qquad \dfrac{12}{9} = \dfrac{16}{??}$
Step 2 Solve the proportion as you did on the previous page.	**Step 2** Solve the proportion as you did on the previous page.
$16 \times 6 = 96$ $96 \div 12 = 8$	$16 \times 9 = 144$ $144 \div 12 = 12$

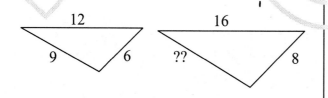 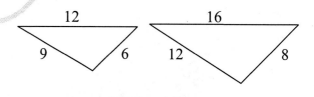

SIMILAR TRIANGLES

Find the missing side from the following similar triangles.

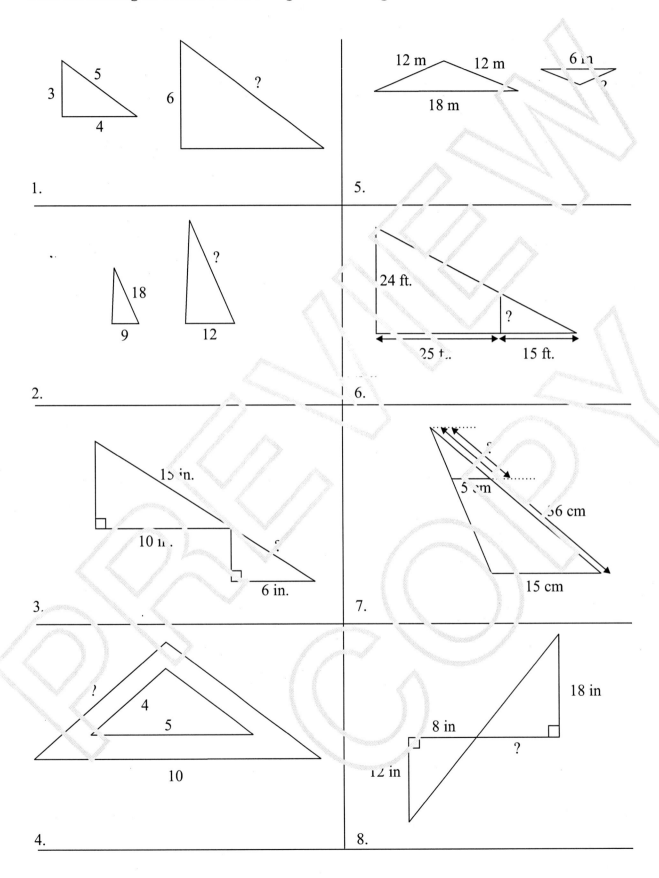

RECOGNIZING SIMILAR TRIANGLES

Similar triangles have sides of different lengths, but the angle measurements are always the same.

EXAMPLE:

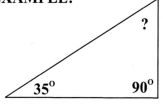
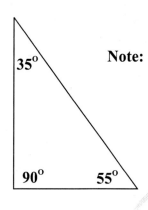

Note: The measures of the three angles inside a triangle always add up to 180°. If you know the measures of two angles, you can subtract the sum from 180° to find the measure of the third angle.

To find the missing angle in the triangle on the far left, add: 35 + 90 = 125. Then 180 - 125 = 55. The missing angle is 55°.

Match the following pairs of similar triangles below.

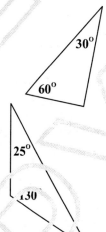
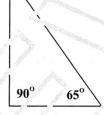
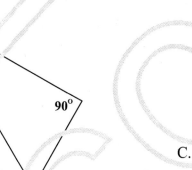
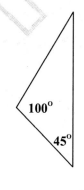
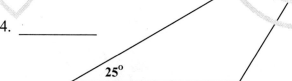
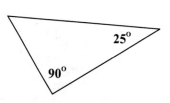

1. _____
2. _____
3. _____
4. _____

A.
B.
C.
D.

253

PYTHAGOREAN THEOREM

Pythagoras was a Greek mathematician and philosopher who lived around 600 B.C. He started a math club among Greek aristocrats called the Pythagoreans. Pythagoras formulated the **Pythagorean Theorem** which states that in a **right triangle**, the sum of the squares of the legs of the triangle are equal to the square of the hypotenuse. Most often you will see this formula written as $a^2 + b^2 = c^2$. **This relationship is only true for right triangles.**

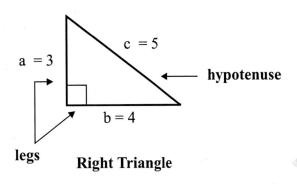

Formula: $a^2 + b^2 = c^2$
$3^2 + 4^2 = 5^2$
$9 + 16 = 25$
$25 = 25$

EXAMPLES:

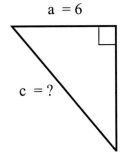

$6^2 + 8^2 = c^2$ ← Square each leg of the triangle and add.
$36 + 64 = c^2$
$100 = c^2$
$\sqrt{100} = \sqrt{c^2}$ ← Find the square root of each side to solve for c.
$10 = c$

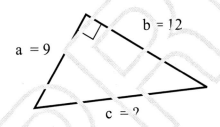

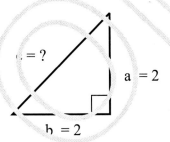

$9^2 + 12^2 = c^2$
$81 + 144 = c^2$
$225 = c^2$
$\sqrt{225} = \sqrt{c^2}$
$15 = c$

$2^2 + 2^2 = c^2$
$4 + 4 = c^2$
$8 = c^2$
$\sqrt{8} = \sqrt{c^2}$
$2.83 = c$

Find the hypotenuse of the following triangles. Round the answers to two decimal places.

1.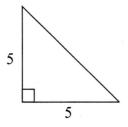

   c = _____

2.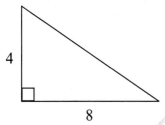

   c = _____

3.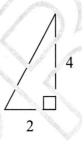

   c = _____

4.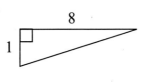

   c = _____

5.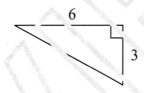

   c = _____

6.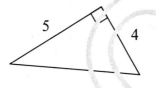

   c = _____

7.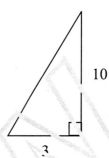

   c = _____

8.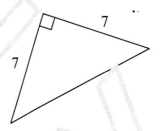

   c = _____

9.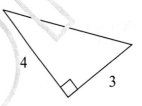

   c = _____

255

CHAPTER REVIEW

Name the following figures.

1.
   _____

2.
   _____

Use the diagram below to answer questions 3 and 4.

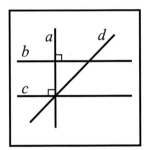

3. Name two lines that are parallel.
   _____

4. Name a vertical line.
   _____

5. Calculate the perimeter of the following figure.

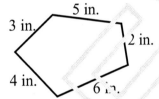

   _____

Calculate the perimeter and area of the following figures.

6.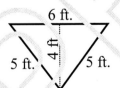
   P = _____
   A = _____

7.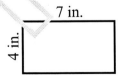
   P = _____
   A = _____

Use the circle diagram below to answer questions 8-10.

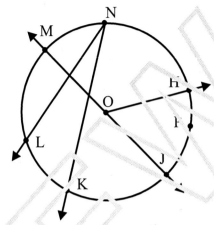

8. Which line segment is a diameter of the circle?
   _____

9. Which angle is an acute central angle?
   _____

10. Name the smallest minor arc that contains point P.
    _____

Calculate the circumference and the area of the following circles.

11. (7 cm)
    Use $\pi = \frac{22}{7}$
    C = _____
    A = _____

12.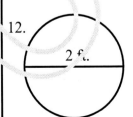
    Use $\pi = 3.14$
    C = _____
    A = _____

13. Use π = 3.14 to find the area of the shaded part. Round your answer to the nearest whole number.

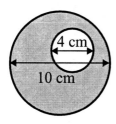

A = _____

14. Use a ruler to measure the dimensions of the following figure in inches. Find the perimeter.

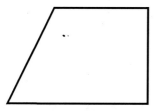

P = _____

15. Use a ruler to measure the dimension of the following figure in centimeters. Find the perimeter and area.

P = _____

A = _____

16. Measure the object below with a ruler, and estimate the area in square centimeters.

Area is about _____

Use the figures below to answer questions 17 and 18.

fig. a   fig. b   fig. c

17. Which two figures above are congruent?

_____

18. Which two figures above are similar?

_____

19. Find the missing side of the triangle below.

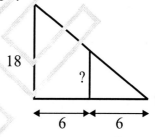

_____

20. The following triangles are similar. Find the missing angle.

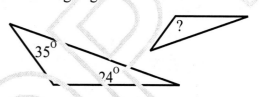

_____

21. Find the missing side. Round your answer to two decimal places.

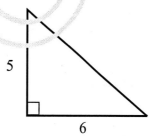

_____

CHAPTER TEST

1. Which of the following is an isosceles triangle?

   A.

   B.

   C.

   D.

2. Which line segments are perpendicular in the figure below?

   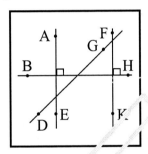

   A. $\overline{AE}$ and $\overline{BH}$
   B. $\overline{GD}$ and $\overline{FK}$
   C. $\overline{AE}$ and $\overline{GD}$
   D. $\overline{BH}$ and $\overline{GD}$

3. Which of the following figures is a parallelogram?

   A.

   B.

   C.

   D.

4. Find the <u>perimeter</u> of the triangle below.

   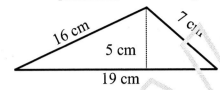

   A. 35 cm
   B. 23 cm
   C. 42 cm
   D. 114 cm

5. What is the perimeter of a hexagon if each side measures 8 inches?

   A. 48 inches
   B. 40 inches
   C. 32 inches
   D. 24 inches

6. Chip wants to put a multicolored racing stripe around his tool box. The tool box measures 30 inches by 12 inches. How many inches of racing stripe will Chip need?

   A. 42 inches
   B. 48 inches
   C. 84 inches
   D. 360 inches

7. Using the formula $A = \frac{1}{2}bh$ for the area of a triangle, find the area of the triangle below.

   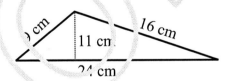

   A. 49 cm^2
   B. 132 cm^2
   C. 264 cm^2
   D. 3,456 cm^2

8. Find the area of the rectangle below.

   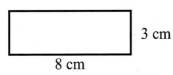

   A. 11 cm²
   B. 22 cm²
   C. 24 cm²
   D. 32 cm²

9. Keri measured her laundry room for new floor covering. She found it measured 4 feet by 8 feet. What is the area of the floor of her laundry room?

   A. 12 ft²
   B. 24 ft²
   C. 32 ft²
   D. 48 ft²

10. What is the area of the following circle? Use $A = \pi r^2$, $\pi = 3.14$.

    A. 12.56 in²
    B. 16 in²
    C. 25.2 in²
    D. 50.24 in²

11. A Ferris wheel has a radius of 14 feet. How far will you travel if you take a ride that goes around six times?

    Formulas: Circumference = $2\pi r$
    Area = πr^2

    Use $\pi = \frac{22}{7}$.

    A. 264 feet
    B. 528 feet
    C. 3,696 feet
    D. 12,936 feet

12. Tara stands in the center of a circle that is 60 inches across. How far is she from the edge of the circle?

    A. about 19 inches
    B. about 10 inches
    C. exactly 30 inches
    D. exactly 60 inches

13. Using a ruler, estimate the area of the following figure in square inches.

    A. 1 in²
    B. 2 in²
    C. 4 in²
    D. 6 in²

14. Cynthia wants to buy a daycare center with the measurements below. How many square feet are in the building?

    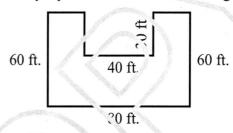

    A. 540 ft²
    B. 480 ft²
    C. 3,600 ft²
    D. 4,800 ft²

15. A bricklayer uses 7 bricks per square foot of surface covered. If he covers an area 28 feet wide and 10 feet high, how many bricks will he use?

    A. 280
    B. 287
    C. 1,960
    D. 2,870

16. The following triangles are similar.

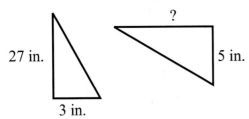

What is the length of the missing side?

A. 45 in.
B. 18 in.
C. 15 in.
D. 10 in.

17. The living room in Ty's house has 168 square feet of floor space. His family is building on an addition to the room that measures 14 feet long and 8 feet wide. What will be the total square feet of the living room with the new addition?

A. 112 ft^2
B. 180 ft^2
C. 270 ft^2
D. 280 ft^2

18. What is the length of line segment $\overline{WY}$?

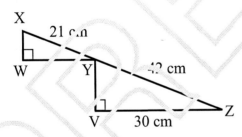

A. 10 cm
B. 15 cm
C. 18 cm
D. 30 cm

19. What is the area of a 5 inch square?

A. 5 in^2
B. 10 in^2
C. 20 in^2
D. 25 in^2

20. On the way to school, Bobby noticed the school bus travels 8 miles north, then turns west and goes 6 miles to get to school. If there were a road that went straight from Bobby's house to school, how long would the road be?

A. 4 miles
B. 9 miles
C. 10 miles
D. 14 miles

21.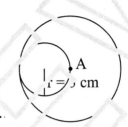

Point A is in the center of the larger circle. The radius of the smaller circle is 3 cm. What is the area of the larger circle? Use 3.14 for π. Area = πr^2.

A. 9.42
B. 18.84
C. 28.26
D. 113.04

22. The area of the shaded region of the rectangle below is 6 square feet. What is the length of the rectangle?

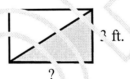

A. 2 ft.
B. 3 ft.
C. 4 ft.
D. 6 ft.

23. What is the circumference of a circle that is 3 feet in diameter? Use C = πd or C = $2\pi r$ and π = 3.14.

A. 4.71 feet
B. 6.00 feet
C. 9.42 feet
D. 18.84 feet

24.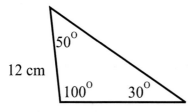

Which triangle is **similar** to the triangle above?

A.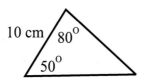

B.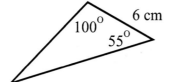

C.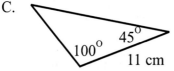

D.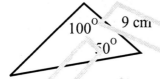

25. Use a ruler to estimate the area of the figure below.

A. 1 square centimeter
B. 2 square centimeters
C. 3 square centimeters
D. 4 square centimeters

26. Harold wants to fence in a rectangular area on all 4 sides. The dimensions of the area are 50 feet by 100 feet. How much fencing should he buy?

A. 150 feet
B. 300 feet
C. 400 feet
D. 5,000 feet

27. The figure below is a circle inscribed in a square. What is the area of the shaded region? The formula for area of a circle is $A = \pi r^2$. Use $\pi = \frac{22}{7}$.

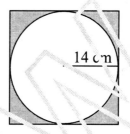

A. 44 cm^2
B. 168 cm^2
C. 196 cm^2
D. 616 cm^2

28. In the diagram below, which arc is a major arc?

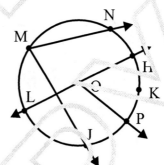

A. $\overset{\frown}{HKP}$
B. $\overset{\frown}{LMN}$
C. $\overset{\frown}{HMP}$
D. $\overset{\frown}{NKJ}$

29. Which figure is a hexagon?

A. (pentagon)

B. (heptagon)

C. (hexagon)

D. (heptagon)

261

Chapter 18

SOLID GEOMETRY

In this chapter, you will learn about the following three-dimensional shapes.

SOLIDS

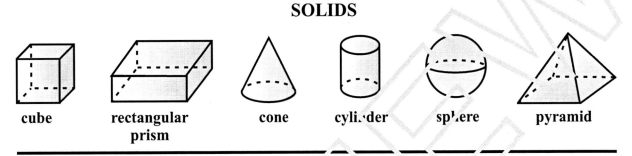

cube   rectangular prism   cone   cylinder   sphere   pyramid

UNDERSTANDING VOLUME

Measurement of <u>volume</u> is expressed in cubic units such as in^3, ft^3, m^3, cm^3, or mm^3. The volume of a solid is the number of cubic units that can be contained in the solid.

First, let's look at rectangular solids.

EXAMPLE:

How many 1 cubic centimeter cubes will it take to fill up the figure below?

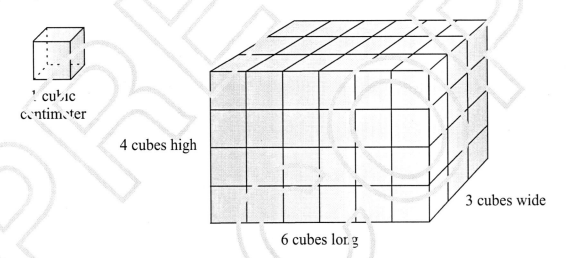

1 cubic centimeter

4 cubes high

3 cubes wide

6 cubes long

To find the volume, you need to multiply the length times the width times the height.

Volume of a rectangular solid = length × width × height   ($V = l\,w\,h$)

$V = 6 \times 3 \times 4 = 72$ cm^3

CUBES

EXAMPLE: Find the volume of the cube pictured at the right.

Each side of a cube has the same measure.

The formula for the volume of a cube is: $V = s^3$ ($s \times s \times s$)

Step 1 Insert measurements from the figure into the formula:

Step 2 Multiply to solve. $5 \times 5 \times 5 = 125 \text{ cm}^3$

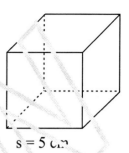

s = 5 cm

Note: Volume is always expressed in cubic units such as in^3, ft^3, cm^3, or m^3.

Answer each of the following questions about cubes.

---

1. If a cube is 3 cm on each edge, what is the volume of the cube?

   _____

2. If the measure of the edge is doubled to 6 cm on each edge, what is the volume of the cube?

   _____

3. How many cubes with edges measuring 3 cm would you need to stack together to make a solid 12 cm cube?

   _____

4. What if the edge of a 3 cm cube is tripled to become 9 cm on each edge? What will the volume be?

   _____

5. What is the volume of a 2 cm cube?

   _____

6. Jerry built a 2 inch cube to hold his marble collection. He wants to build a cube with a volume 8 times larger. How much will each edge measure?

   _____

Find the volume of the following cubes.

7.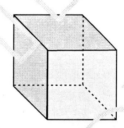

   s = 7 in.   V = _____

8.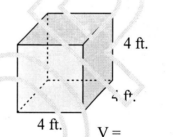

   4 ft.   4 ft.   4 ft.   V = _____

9. 12 inches = 1 foot

   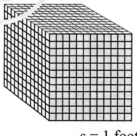

   s = 1 foot

   How many cubic inches are in a cubic foot?

   _____

Find the volume of the following rectangular prisms (boxes) and cubes.

1.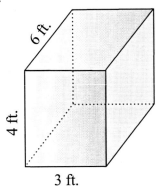

    V = _____

4.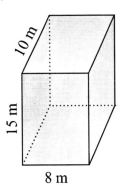

    V = _____

7.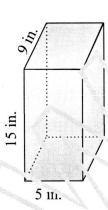

    V = _____

2.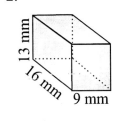

    V = _____

5.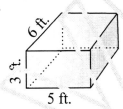

    V = _____

8.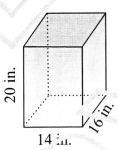

    V = _____

3.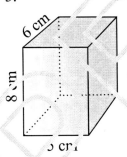

    V = _____

6.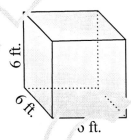

    V = _____

9.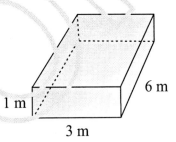

    V = _____

VOLUME OF SPHERES, CONES, CYLINDERS, AND PYRAMIDS

To find the volume of a solid, substitute the measurements given for the solid in the correct formula and solve. Volumes are expressed in cubic units such as ft^3, in^3, cm^3.

Sphere
$V = \frac{4}{3}\pi r^3$

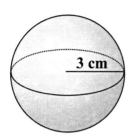

$V = \frac{4}{3}\pi r^3 \quad \pi = 3.14$
$V = \frac{4}{3} \times 3.14 \times 27$
$V = 113.04 \text{ cm}^3$

Cone
$V = \frac{1}{3}\pi r^2 h$

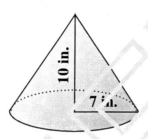

$V = \frac{1}{3}\pi r^2 h \quad \pi = 3.14$
$V = \frac{1}{3} \times 3.14 \times 49 \times 10$
$V = 512.87 \text{ in}^3$

Cylinder
$V = \pi r^2 h$

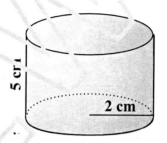

$V = \pi r^2 h \quad \pi = \frac{22}{7}$
$V = \frac{22}{7} \times 4 \times 5$
$V = 62.86 \text{ cm}^3$

Pyramids

$V = \frac{1}{3}bh \quad b = \text{area of rectangular base}$

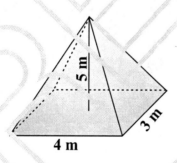

$V = \frac{1}{3}bh \quad b = l \times w$
$V = \frac{1}{3} \times 4 \times 3 \times 5$
$V = 20 \text{ m}^3$

$V = \frac{1}{3}bh \quad b = \text{area of triangular base}$

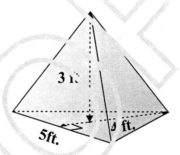

$b = \frac{1}{2} \times 5 \times 4 = 10 \text{ ft}^2$
$V = \frac{1}{3} \times 10 \times 3$
$V = 10 \text{ ft}^3$

Find the Volume of the following shapes. Use π = 3.14.

1. 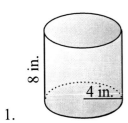 V = _____

7. 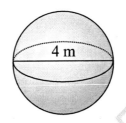 V = _____

2. 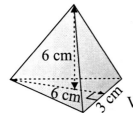 V = _____

8. 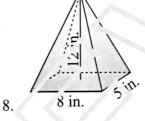 V = _____

3. 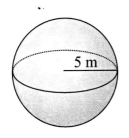 V = _____

9. 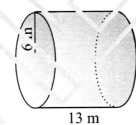 V = _____

4. 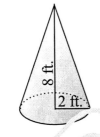 V = _____

10. 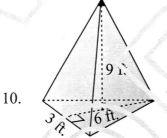 V = _____

5. 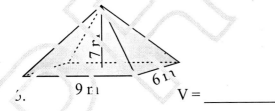 V = _____

11. 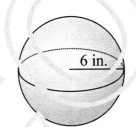 V = _____

6. 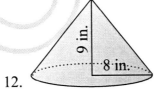 V = _____

12. V = _____

ESTIMATING VOLUME

Measure the following objects with your ruler and estimate the volume.

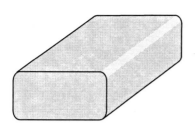

1. Volume is about _____ in³

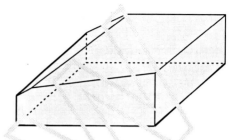

4. Volume is about _____ cm³

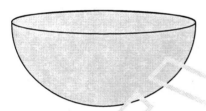

2. Volume is about _____ in³

5. Volume is about _____ cm³.

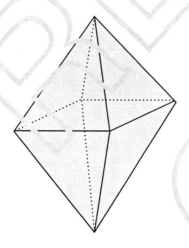

3. Volume is about _____ in³

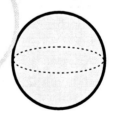

6. Volume is about _____ in³

Copyright © American Book Company

267

SURFACE AREA

The surface area of a solid is the total area of all the sides of a solid.

CUBE

There are six sides on a cube. To find the surface area of a cube, find the area of one side and multiply by 6.

Area of each side of the cube:
$3 \times 3 = 9$ cm^2

Total surface area: $9 \times 6 = 54$ cm^2

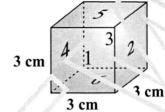

RECTANGULAR PRISM

There are 6 sides on a rectangular prism. To find the surface area, add the areas of the six rectangular sides.

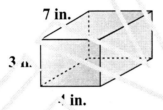

Top and Bottom

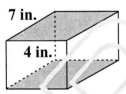

Area of top side:
$7 \times 4 = 28$ in.^{2}
Area of top and bottom:
$28 \times 2 = 56$ in.^{2}

Front and Back

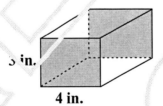

Area of front:
$3 \times 4 = 12$ in.^{2}
Area of front and back:
$12 \times 2 = 24$ in.^{2}

Left and Right

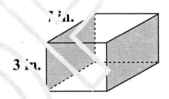

Area of left side:
$3 \times 7 = 21$ in.^{2}
Area of left and right:
$21 \times 2 = 42$ in^2

Total surface area: $56 + 24 + 42 = 122$ in^2

Find the surface area of the following cubes or prisms.

1.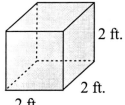

   SA = _____

2.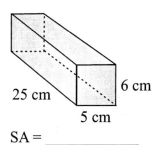

   SA = _____

3.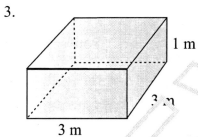

   SA = _____

4.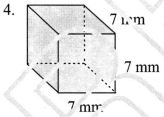

   SA = _____

5.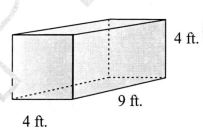

   SA = _____

6.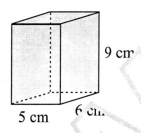

   SA = _____

7.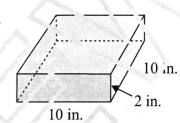

   SA = _____

8.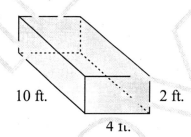

   SA = _____

9.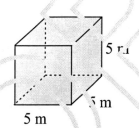

   SA = _____

10.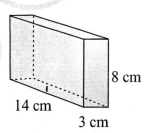

    SA = _____

PYRAMID

The pyramid below is made of a square base with 4 triangles on the sides.

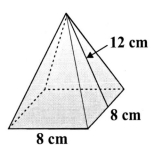
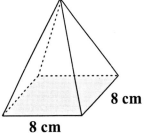
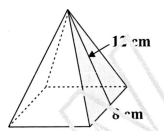

Area of square base:
$A = l \times w$
$A = 8 \times 8 = 64 \text{ cm}^2$

Area of sides.
Area of 1 side = $\frac{1}{2}bh$
$A = \frac{1}{2} \times 8 \times 12 = 48 \text{ cm}^2$
Area of 4 sides = $48 \times 4 = 192 \text{ cm}^2$

Total surface area: $64 + 192 = 256 \text{ cm}^2$

Find the surface area of the following pyramids

1. (3 ft., 2 ft., 2 ft.)

SA = _____

2. (12 mm, 6 mm, 6 mm)

SA = _____

3. (15 m, 10 m, 10 m)

SA = _____

4. 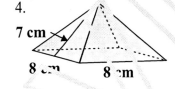 (7 cm, 8 cm, 8 cm)

SA = _____

5. (4 m, 3 m, 3 m)

SA = _____

6. 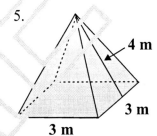 (10 in., 9 in., 9 in.)

SA = _____

7. (9 m, 4 m, 4 m)

SA = _____

8. (10 in., 5 in., 5 in.)

SA = _____

9. 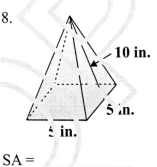 (7 ft., 7 ft., 7 ft.)

SA = _____

CYLINDER

If the side of a cylinder were slit from top to bottom and laid flat, its shape would be a rectangle. The length of the rectangle is the same as the circumference of the circle that is the base of the cylinder. The width of the rectangle is the height of the cylinder.

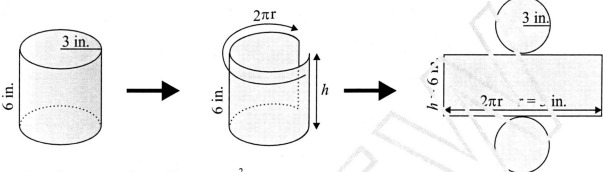

Total Surface Area of a Cylinder = $2\pi r^2 + 2\pi rh$

Area of top and bottom:
Area of a circle = πr^2
Area of top = $3.14 \times 3^2 = 28.26$ in.^{2}
Area of top and bottom = $2 \times 28.26 = 56.52$ in.^{2}

Area of side:
Area of rectangle = $l \times h$
$l = 2\pi r = 2 \times 3.14 \times 3 = 18.84$ in.
Area of rectangle = $18.84 \times 6 = 113.04$ in.^{2}

Total surface area = $56.52 + 113.04 = 169.56$ in^2

Find the surface area of the following cylinders. Use $\pi = 3.14$

1.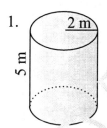

SA = _____

4.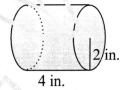

SA = _____

7.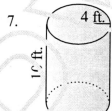

SA = _____

2.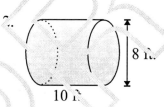

SA = _____

5.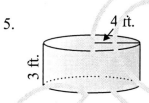

SA = _____

8.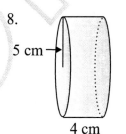

SA = _____

3.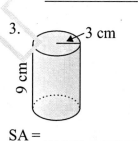

SA = _____

6.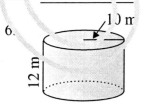

SA = _____

9.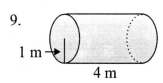

SA = _____

SOLID GEOMETRY WORD PROBLEMS

1. If an Egyptian pyramid has a square base that measures 500 yards by 500 yards, and the pyramid stands 300 yards tall, what would be the volume of the pyramid? Use the formula for volume of a pyramid, $V = \frac{1}{3}bh$ where b is the area of the base.

    V = _____

2. Robert is using a cylindrical barrel filled with water to flatten the sod in his yard. The circular ends have a radius of 1 foot. The barrel is 3 feet wide. How much water will the barrel hold? The formula for volume of a cylinder is $V = \pi r^2 h$. Use $\pi = 3.14$.

    V = _____

3. If a basketball measures 24 centimeters in diameter, what volume of air will it hold? The formula for volume of a sphere is $V = \frac{4}{3}\pi r^3$. Use $\pi = 3.14$.

    V = _____

4. What is the volume of a cone that is 2 inches in diameter and 5 inches tall? The formula for volume of a cone is $V = \frac{1}{3}\pi r^2 h$. Use $\pi = 3.14$.

    V = _____

5. Kelly has a rectangular fish aquarium that measures 24 inches wide, 12 inches deep, and 18 inches tall. What is the maximum amount of water that the aquarium will hold?

    V = _____

6. Jenny has a rectangular box that she wants to cover in decorative contact paper. The box is 10 cm long, 5 cm wide, and 5 cm high. How much paper will she need to cover all 6 sides?

    SA = _____

7. Gasco needs to construct a cylindrical, steel gas tank that measures 6 feet in diameter and is 8 feet long. How many square feet of steel will be needed to construct the tank? Use the following formulas as needed: $A = l \times w$, $A = \pi r^2$, $C = 2\pi r$. Use $\pi = 3.14$.

    SA = _____

8. Craig wants to build a miniature replica of San Francisco's Transamerica Pyramid out of glass. His replica will have a square base that measures 6 cm by 6 cm. The 4 triangular sides will be 6 cm wide and 60 cm tall. How many square centimeters of glass will he need to build his replica? Use the following formulas as needed: $A = l \times w$ and $A = \frac{1}{2}bh$.

    SA = _____

9. Jeff built a wooden, cubic toy box for his son. Each side of the box measured 2 feet. How many square feet of wood did he use to build the toy box? How many cubic feet of toys will the box hold?

    SA = _____

    V = _____

CHAPTER REVIEW

Find the volume and/or the surface area of the following solids.

1.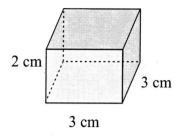

   V = _____

   SA = _____

2.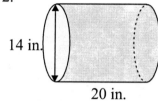

   $V = \pi r^2 h$
   $SA = 2\pi r^2 + 2\pi r h$

   Use $\pi = \frac{22}{7}$

   V = _____

   SA = _____

3.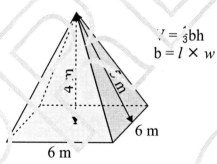

   $V = \frac{1}{3}bh$
   $b = l \times w$

   V = _____

   SA = _____

4.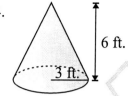

   $V = \frac{1}{3}\pi r^2 h$

   Use $\pi = 3.14$

   V = _____

5.

   $V = \frac{1}{3}bh$
   b = area of the triangular base

   V = _____

6.

   $V = \frac{4}{3}\pi r^3$

   Use $\pi = \frac{22}{7}$

   V = _____

7. Use a ruler to estimate the volume of the following figure in cubic centimeters.

   V = _____

CHAPTER TEST

1. Which of the following figures is a cone?

   A.

   B. (cube)

   C. (cylinder)

   D. (pyramid)

2. What is the volume of the following box?

   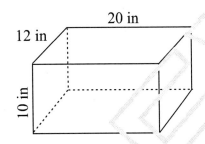

   A. 240 in^2
   B. 240 in^3
   C. 2,400 in^2
   D. 2,400 in^3

3. What is the volume of a sphere with a radius of 4 inches? Round your answer to the nearest hundredth.

   Use the formula $V = \frac{4}{3}\pi r^3$ $\pi = 3.14$

   A. 66.99 in^3
   B. 200.96 in^3
   C. 267.95 in^3
   D. 803.84 in^3

4. Find the volume of the following pyramid.

   Use the formula $V = \frac{lwh}{3}$

   l = length
   w = width
   h = height

   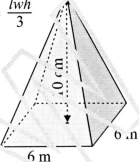

   A. 22 cm^3
   B. 120 cm^3
   C. 360 cm^3
   D. 1,080 cm^3

5. Find the volume of the cone.

   Use the formula $V = \frac{1}{3}\pi r^2 h$ $\pi = \frac{22}{7}$

   A. 88 cm^3
   B. 176 cm^3
   C. 528 cm^3
   D. 2,112 cm^3

6. What is the volume of the following oil tank? Round your answer to the nearest hundredth.

   Use the formula $V = \pi r^2 h$ $\pi = 3.14$

   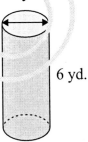

   A. 18.84 yd^3
   B. 37.68 yd^3
   C. 44.48 yd^3
   D. 75.36 yd^3

7. What is the volume of a cube that is 7 inches on each edge?

   A. 42 in³
   B. 49 in³
   C. 84 in³
   D. 343 in³

8. Sabina wants to cover a box on all six sides with white satin. The box is 4"×6"×10". If she glues the fabric on so it does not overlap, how many square inches of fabric will she use?

   A. 120 in²
   B. 124 in²
   C. 240 in²
   D. 248 in²

9. What is the surface area of a cylinder with an 8 inch diameter and a height of 3 inches?

   Use the formula $SA = 2\pi r^2 + 2\pi rh$
   Use $\pi = 3.14$

   A. 100.48 in²
   B. 50.24 in²
   C. 75.36 in²
   D. 175.84 in²

10. A rectangular box and a rectangular pyramid have the same dimensions for their bases and heights. How does the volume of the box compare to the volume of the pyramid?

    A. The volumes are the same.
    B. The volume of the box is twice the volume of the pyramid.
    C. The volume of the box is three times as large as the pyramid.
    D. The volume of the box is four times as large as the pyramid.

11. Measure the height, length, and width of the object below. What is the best estimate for its volume?

    A. 2 cm³
    B. 4 cm³
    C. 10 cm³
    D. 40 cm³

12. Micki built a 4-sided pyramid with the dimensions below. What is the total surface area of the pyramid?

    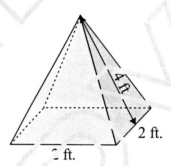

    A. 4 ft²
    B. 16 ft²
    C. 20 ft²
    D. 36 ft²

13. Jack is going to paint the ceiling and four walls of a room that is 10 feet wide, 12 feet long, and 10 feet from floor to ceiling. How many square feet will he paint?

    A. 120 square feet
    B. 560 square feet
    C. 680 square feet
    D. 1,200 square feet

Chapter 19

GRAPHING

GRAPHING ON A NUMBER LINE

Number lines allow you to graph values of positive and negative numbers as well as zero. Any real number, whether it is a fraction, decimal, or integer can be plotted on a number line. Number lines can be horizontal or vertical as you saw in Chapter 12 on Integers. The examples below illustrate how to plot different types of numbers on a number line.

GRAPHING FRACTIONAL VALUES

EXAMPLE 1: What number does point A represent on the number line below?

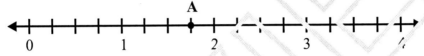

Step 1: Point A is between the numbers 1 and 2, so it is greater than 1 but less than 2. We can express the value of A as a fractional value that falls between 1 and 2. To do so, copy the integer that point A falls between which is closer to zero on the number line. In this case, copy the 1 because 1 is closer to zero on the number line than the 2.

Step 2: Count the number of spaces between each integer. In this case, there are 4 spaces between the 1 and the 2. Put this number as the bottom number in your fraction.

Step 3: Count the number of spaces between the 1 and point A. Point A is 3 spaces away from the number 1. Put this number as the top number in your fraction.

Point A is at $1\frac{3}{4}$

- The integer that point A falls between that is closest to 0
- The number of spaces between 1 and A
- The number of spaces between 1 and 2

EXAMPLE 2: What number does point B represent on the number line below?

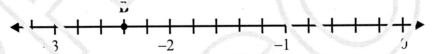

Step 1: Point B is between −2 and −3. Again, we can express the value of B as a fraction that falls between −2 and −3. Copy the integer that point B falls between which is closer to zero. The −2 is closer to zero than −3, so copy −2.

Step 2: In this example, there are 5 spaces between each integer. Five will be the bottom number in the fraction.

Step 3: There are 2 spaces between −2 and point B. Two will be the top number in the fraction.

Point B is at $-2\frac{2}{5}$

Determine and record the value of each point on the number lines below.

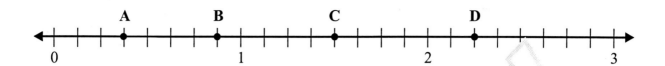

1.   A = ____   B = ____   C = ____   D = ____

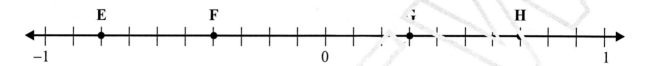

2.   E = ____   F = ____   G = ____   H = ____

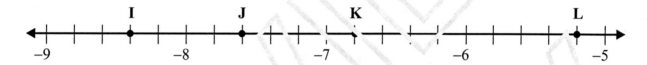

3.   I = ____   J = ____   K = ____   L = ____

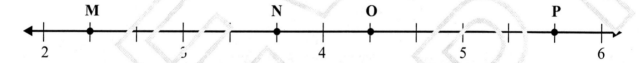

4.   M = ____   N = ____   O = ____   P = ____

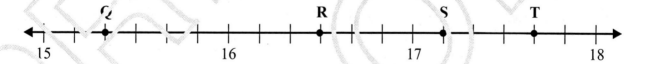

5.   Q = ____   R = ____   S = ____   T = ____

6.   U = ____   V = ____   W = ____   X = ____

RECOGNIZING IMPROPER FRACTIONS, DECIMALS, AND SQUARE ROOT VALUES ON A NUMBER LINE

Improper fractions, decimal values and square root values can also be plotted on a number line. Study the examples below.

EXAMPLE 1: Where would $\frac{4}{3}$ fall on the number line below?

Step 1: Convert the improper fraction to a mixed number. $\frac{4}{3} = 1\frac{1}{3}$

Step 2: $1\frac{1}{3}$ is $\frac{1}{3}$ of the distance between the numbers 1 and 2. Estimate this distance by dividing the distance between points 1 and 2 into thirds. Plot the point at the first division.

EXAMPLE 2: Plot the value of −1.75 on the number line below.

Step 1: Convert the value −1.75 to a mixed fraction. $-1.75 = -1\frac{3}{4}$

Step 2: $-1\frac{3}{4}$ is $\frac{3}{4}$ of the distance between the numbers −1 and −2. Estimate this distance by dividing the distance between points −1 and −2 into fourths. Plot the point at the third division.

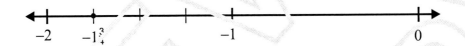

EXAMPLE 3: Plot the value of $\sqrt{3}$ on the number line below.

Step 1: Estimate the value of $\sqrt{3}$ by using the square root of values that you know. $\sqrt{1} = 1$ and $\sqrt{4} = 2$, so the value of $\sqrt{3}$ is going to be between 1 and 2.

Step 2: To estimate a little closer, try squaring 1.5. $1.5 \times 1.5 = 2.25$, so $\sqrt{3}$ has to be greater than 1.5. If you do further trial and error calculations, you will find that $\sqrt{3}$ is greater than 1.7 ($1.7 \times 1.7 = 2.89$) but less than 1.8 ($1.8 \times 1.8 = 3.24$).

Step 3: Plot $\sqrt{3}$ around 1.75.

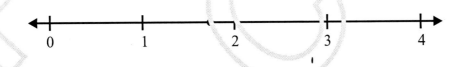

Plot and label the following values on the number lines given below.

1. A = $\frac{5}{4}$ B = $\frac{12}{5}$ C = $\frac{2}{3}$ D = $-\frac{3}{2}$

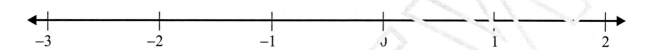

2. E = 1.4   F = –2.25   G = –0.6   H = 0.625

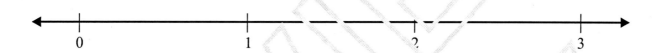

3. I = $\sqrt{2}$ J = $\sqrt{5}$ K = $\sqrt{3}$ L = $\sqrt{8}$

Match the correct value for each point on the number lines below

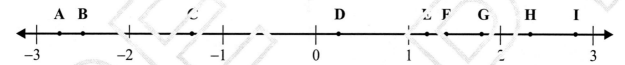

4. 1.8 = ____   7. $-\frac{5}{2}$ = ____   10. $\sqrt{8}$ = ____

5. $\frac{7}{3}$ = ____   8. –2.75 = ____   11. $\frac{6}{5}$ = ____

6. $\sqrt{2}$ = ____   9. $-\frac{4}{3}$ = ____   12. 0.25 = ____

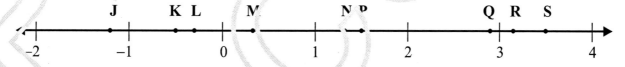

13. $\sqrt{12}$ = ____   16. $-\frac{1}{3}$ = ____   19. $-\frac{6}{5}$ = ____

14. –0.5 = ____   17. 1.5 = ____   20. $\sqrt{10}$ = ____

15. $\frac{5}{4}$ = ____   18. –0.3 = ____   21. 2.9 = ____

PLOTTING POINTS ON A VERTICAL NUMBER LINE

Number lines can also be drawn up and down (**vertical**) instead of across the page (**horizontal**). You plot points on a vertical number line the same way as you do on a horizontal number line.

Record the value represented by each point on the number lines below.

1. A = _____
2. B = _____
3. C = _____
4. D = _____
5. E = _____
6. F = _____
7. G = _____
8. H = _____

9. I = _____
10. J = _____
11. K = _____
12. L = _____
13. M = _____
14. N = _____
15. P = _____
16. Q = _____

17. Q = _____
18. R = _____
19. S = _____
20. T = _____
21. U = _____
22. W = _____
23. X = _____
24. Y = _____

25. A = _____
26. B = _____
27. C = _____
28. D = _____
29. E = _____
30. G = _____
31. H = _____
32. I = _____

CARTESIAN COORDINATES

A number line allows you to graph points with only one value. A **Cartesian coordinate plane** allows you to graph points with two values. A Cartesian coordinate plane is made up of two number lines. The horizontal number line is called the **x-axis** and the vertical number line is called the **y-axis**. The point where the x and y axes intersect is called the **origin**. The x and y axes separate the Cartesian coordinate plane into four quadrants that are labeled I, II, III, and IV. The quadrants are labeled and explained on the graph below. Each point graphed on the plane is designated by an **ordered pair** of coordinates. For example, $(2, -1)$ is an ordered pair of coordinates designated by **point B** on the plane below. The first number, 2, tells you to go over positive two on the x-axis. The -1 tells you to then go down negative one on the y-axis.

Remember: The first number always tells you how far to go right or left of 0, and the second number always tells you how far to go up or down from 0.

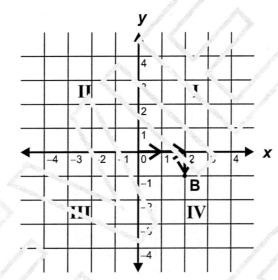

Quadrant II:
The x-coordinate is negative, and the y-coordinate is positive $(-, +)$.

Quadrant III:
Both coordinates in the ordered pair are negative $(-, -)$.

Quadrant I:
Both coordinates in the ordered pair are positive $(+, +)$.

Quadrant IV:
The x-coordinate is positive and the y-coordinate is negative $(+, -)$.

Plot and label the following points on the Cartesian coordinate plane provided.

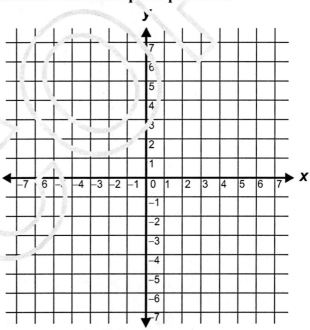

A. (2, 4)
B. (-1, 5)
C. (3, -4)
D. (-5, -2)
E. (5, 3)
F. (-7, -6)
G. (-2, 5)
H. (6, -1)
I. (4, -7)
J. (6, 2)

K. (-1, -1)
L. (3, -3)
M. (5, 5)
N. (-2, -2)
O. (0, 0)
P. (0, 4)
Q. (2, 0)
R. (-4, 0)
S. (0, -2)
T. (5, 1)

IDENTIFYING ORDERED PAIRS

When identifying **ordered pairs**, count how far left or right of 0 to find the *x*-coordinate and then how far up or down from 0 to find the *y*-coordinate.

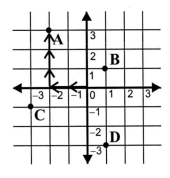

Point A: Left (negative) two and up (positive) three = (−2, 3) in quadrant II

Point B: Right (positive) one and up (positive) one = (1, 1) in quadrant I

Point C: Left (negative) three and down (negative) one = (−3, −1) in quadrant III

Point D: Right (positive) one and down (negative) three = (1, −3) in quadrant IV

Fill in the ordered pair for each point and tell which quadrant it is in.

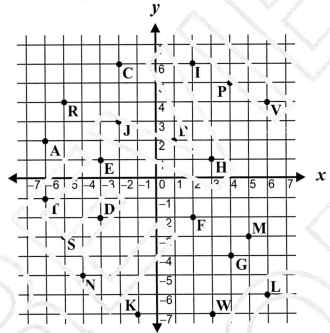

1. point A = (,) quadrant _____
2. point B = (,) quadrant _____
3. point C = (,) quadrant _____
4. point D = (,) quadrant _____
5. point E = (,) quadrant _____
6. point F = (,) quadrant _____
7. point G = (,) quadrant _____
8. point H = (,) quadrant _____
9. point I = (,) quadrant _____
10. point J = (,) quadrant _____
11. point K = (,) quadrant _____
12. point L = (,) quadrant _____
13. point M = (,) quadrant _____
14. point N = (,) quadrant _____
15. point P = (,) quadrant _____
16. point R = (,) quadrant _____
17. point S = (,) quadrant _____
18. point T = (,) quadrant _____
19. point V = (,) quadrant _____
20. point W = (,) quadrant _____

Sometimes, points on a coordinate plane fall on the *x* or *y* axis. If a point falls on the *x*-axis, then the second number of the ordered pair is 0. If a point falls on the *y*-axis, the first number of the ordered pair is 0.

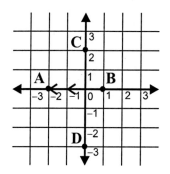

Point A: Left (negative) two and up zero = (−2, 0)
Point B: Right (positive) one and up zero = (1, 0)
Point C: Left/right zero and up (positive) two = (0, 2)
Point D: Left/right zero and down (negative) three = (0, −3)

Fill in the ordered pair for each point.

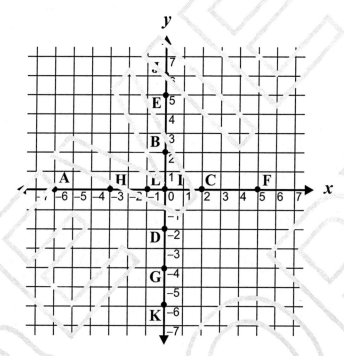

1. point A = (,)
2. point B = (,)
3. point C = (,)
4. point D = (,)
5. point E = (,)
6. point F = (,)
7. point G = (,)
8. point H = (,)
9. point I = (,)
10. point J = (,)
11. point K = (,)
12. point L = (,)

DRAWING GEOMETRIC FIGURES ON A CARTESIAN COORDINATE PLANE

You can use a **Cartesian coordinate plane** to draw geometric figures by plotting **vertices** and connecting them with line segments.

EXAMPLE 1: What are the coordinates of each vertex of quadrilateral ABCD below?

Step 1: To find the coordinates of point A, count over −3 on the x-axis and up 1 on the y-axis.
Point A = (−3, 1)

Step 2: The coordinates of point B are located to the right two units on the x-axis and up 3 units on the y-axis.
Point B = (2, 3)

Step 3: Point C is located 4 units to the right on the x-axis and down −3 on the y-axis.
Point C = (4, −3)

Step 4: Point D is −4 units left on the x-axis and down −4 units on the y-axis.
Point D = (−4, −4)

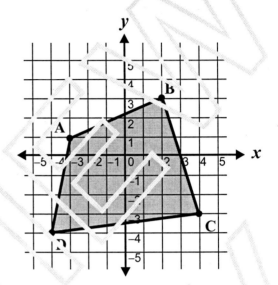

EXAMPLE 2: Plot the following points. Then construct and identify the geometric figure that you plotted.

A = (−2, −5), B = (−2, 1), C = (3, 1), D = (3, −5)

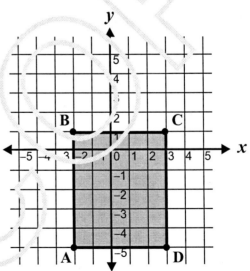

Figure ABCD is a rectangle.

Find the coordinates of the geometric figures graphed below.

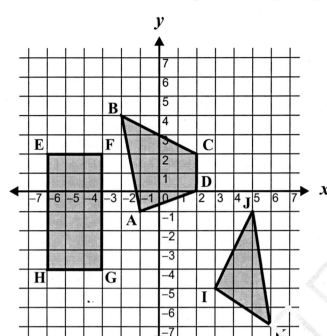

1. quadrilateral ABCD

   A = _____
   B = _____
   C = _____
   D = _____

2. rectangle EFGH

   E = _____
   F = _____
   G = _____
   H = _____

3. triangle IJK

   I = _____
   J = _____
   K = _____

---

4. parallelogram LMNO

   L = _____
   M = _____
   N = _____
   O = _____

5. right triangle PQR

   P = _____
   Q = _____
   R = _____

6. pentagon STVXY

   S = _____
   T = _____
   V = _____
   X = _____
   Y = _____

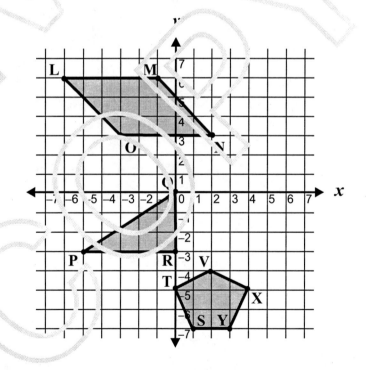

Plot and label the following points. Then construct and identify the geometric figure you plotted. Question 1 is done for you.

Figure

1. Point A = (−1, −1)
   Point B = (−1, 2)
   Point C = (2, 2)
   Point D = (2, −1)   square

2. Point E = (3, −2)
   Point F = (5, 1)
   Point G = (7, −2)   _____

3. Point H = (−4, 0)
   Point I = (−6, 0)
   Point J = (−4, 4)
   Point K = (−2, 4)   _____

4. Point L = (−1, −3)
   Point M = (4, −6)
   Point N = (−1, −6)   _____

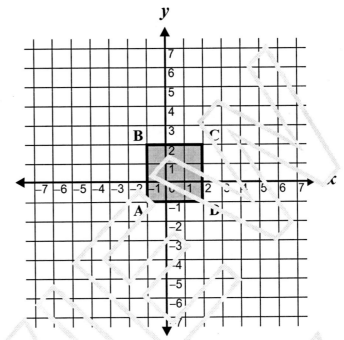

figure

5. Point A = (−2, 3)
   Point B = (−3, 5)
   Point C = (−1, 5)
   Point D = (1, 5)
   Point E = (0, 3)   _____

6. Point F = (−1, −3)
   Point G = (−3, −5)
   Point H = (−1, −7)
   Point I = (1, −5)   _____

7. Point J = (−1, 2)
   Point K = (−1, −1)
   Point L = (3, −2)   _____

8. Point M = (6, 2)
   Point N = (6, −4)
   Point O = (4, −4)
   Point P = (4, 2)   _____

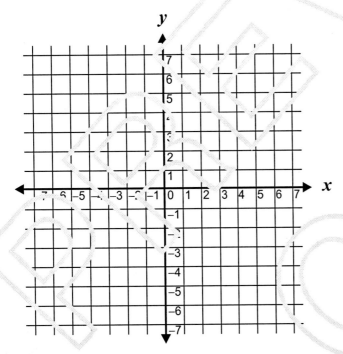

286   Copyright © American Book Company

CHAPTER REVIEW

1.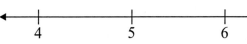

   Plot and label $5\frac{3}{5}$ on the number line above.

2.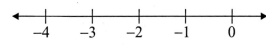

   Plot and label $-3\frac{1}{2}$ on the number line above.

3.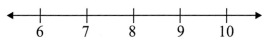

   Plot and label $\sqrt{52}$ on the number line above.

4.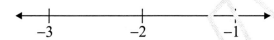

   Plot and label -2.3 on the number line above.

Record the value represented by the point on the number line for questions 5 - 10.

5. A _____

6. B _____

7. C _____

8. D _____

9. E _____

10. F _____

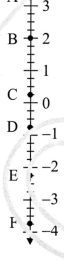

Record the coordinates and quadrants of the following points.

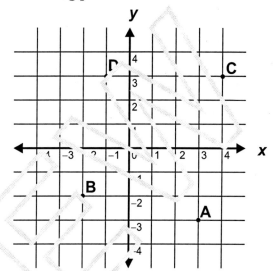

        Coordinates   Quadrant

11. A = _____ _____

12. B = _____ _____

13. C = _____ _____

14. D = _____ _____

On the same plane above label these additional coordinates.

15. E = (0, -3)

16. F = (-3, -1)

17. G = (4, 0)

18. H = (2, 2)

Answer the following questions.

19. In which quadrant does the point (2, 3) lie? _____

20. In which quadrant does the point (-5, -2) lie? _____

CHAPTER TEST

1.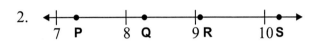

   Which point on the number line above represents $5\frac{3}{4}$?

   A. Point Q
   B. Point R
   C. Point S
   D. Point T

2.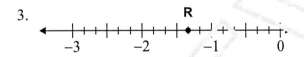

   Where is $\sqrt{83}$ located on the number line above?

   A. Point P
   B. Point Q
   C. Point R
   D. Point S

3.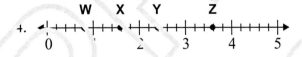

   What number corresponds to point R on the number line above?

   A. $-2\frac{2}{3}$
   B. $-1\frac{2}{3}$
   C. $-1\frac{1}{3}$
   D. $-1\frac{1}{2}$

4.

   Which point on the number line above represents $\frac{8}{5}$?

   A. W
   B. X
   C. Y
   D. Z

5.

   Where is $-2\frac{3}{4}$ located on the number line above?

   A. Point R
   B. Point S
   C. Point T
   D. Point U

6.

   What is the value of point F?

   A. 8.25
   B. 8.3
   C. 8.4
   D. 8.6

7. Which quadrant contains the point $(-2, -4)$?

   A. I
   B. II
   C. III
   D. IV

8. Which quadrant has ordered pairs with positive numbers for both coordinates?

   A. I
   B. II
   C. III
   D. IV

9.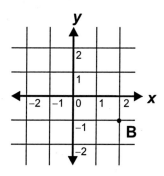

What are the coordinates of point B above?

A. (2, –1)
B. (–1, 2)
C. (–2, 1)
D. (1, –2)

10.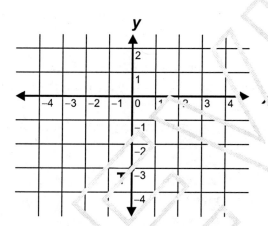

What are the coordinates of Point J on the graph above?

A. (0, 3)
B. (0, –3)
C. (–3, 0)
D. (3, 0)

11. The coordinates of a figure are (–3, –3), (–3, 6), (2, 6), and (2, –3). What is the shape of the figure?

A. square
B. rectangle
C. trapezoid
D. parallelogram

12.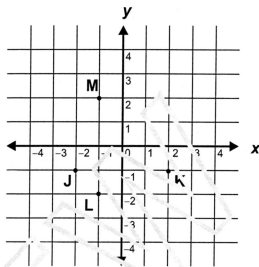

Which point on the graph above has the coordinates (–1, 2)?

A. Point J
B. Point K
C. Point L
D. Point M

13.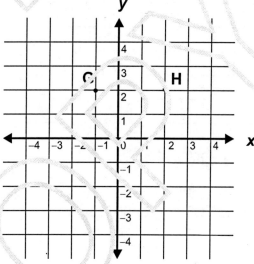

Randy plotted points G and H on the Cartesian coordinate graph above. Where could he plot points J and K if he wants to form a square GHJK?

A. J at (2, 4) and K at (–1, 4)
B. J at (3, 0) and K at (0, 0)
C. J at (2, 0) and K at (–1, 0)
D. J at (2, –1) and K at (–1, –1)

Chapter 20

TRANSFORMATIONS AND SYMMETRY

Transformations are geometric figures that have been changed by **reflection, rotation, and translation**.

REFLECTIONS

A **reflection** of a geometric figure is a mirror image of the object. Placing a mirror on the **line of reflection** will give you the position of the reflected image. On paper, folding an image across the line of reflection will give you the position of the reflected image.

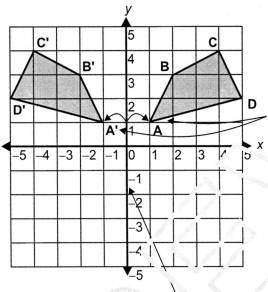

line of reflection: y-axis

Quadrilateral ABCD is reflected across the y-axis to form quadrilateral A'B'C'D'. The y-axis is the line of reflection. Point A' is the reflection of point A, point B' corresponds to point B, C' to C, and D' to D.

Point A is +1 space from the y-axis. Point A's mirror image, point A', is −1 space from the y-axis.

Point B is +2 spaces from the y-axis. Point B' is −2 spaces from the y-axis.

Point C is +4 spaces from the y-axis and point C' is −4 spaces from the y-axis.

Point D is +5 spaces from the y-axis and point D' is −5 spaces from the y-axis.

Triangle FGH is reflected across the x-axis to form triangle F'G'H'. The x-axis is the line of reflection. Point F' reflects point F. Point G' corresponds to point G, and H' mirrors H.

Point F is +3 space from the x-axis. Likewise, point F' is −3 space from the x-axis.

Point G is +1 spaces from the x-axis, and point G' is −1 spaces from the x-axis.

Point H is 0 spaces from the x-axis, so point H' is also 0 spaces from the x-axis.

line of reflection: x-axis

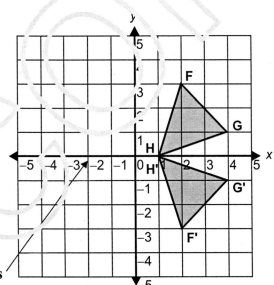

290 Copyright © American Book Company

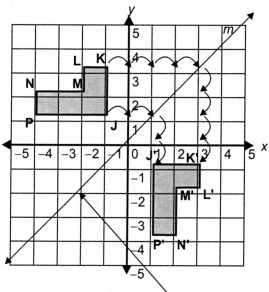

Figure JKLMNP is reflected across line *m* to form figure J'K'L'M'N'P'. Line *m* is at a 45° angle. Point J corresponds to J', K to K', L to L', M to M', N to N' and P to P'. Line m is the line of reflection. **Pay close attention to how to determine the mirror image of figure JKLMNP across line *m* described below. This method only works when the line of reflection is at a 45° angle.**

Point J is 2 spaces over from line *m*, so J' must be 2 spaces down from line *m*.

Point K is 4 spaces over from line m, so K' is 4 spaces down from line *m*, and so on.

line of reflection: line *m*

Draw the following reflections, and record the new coordinates of the reflection. The first problem is done for you.

1. Reflect figure ABC across the x-axis. Label vertices A'B'C' so that point A' is the reflection of point A, B' is the reflection of B, and C' is the reflection of C.

    A' = (–4, –2)   B' = (–2, –4)   C' = (0, –4)

2. Reflect figure ABC across the y-axis. Label vertices A"B"C" so that point A" is the reflection of point A, B" is the reflection of B, and C" is the reflection of C.

    A" = _____   B" = _____   C" = _____

3. Reflect figure ABC across line p. Label vertices A'"B'"C'" so that point A'" is the reflection of point A, B'" is the reflection of B, and C'" is the reflection of C.

    A'" = _____   B'" = _____   C'" = _____

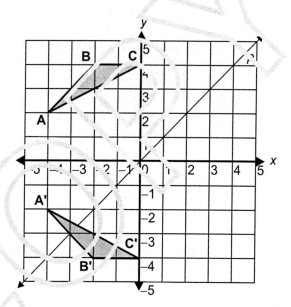

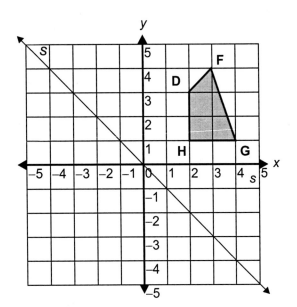

4. Reflect figure DFGH across the y-axis. Label vertices D'F'G'H' so that point D' is the reflection of point D, F' is the reflection of F, G' is the reflection of G, and H' is the reflection of H.

D' = _____ G' = _____

F' = _____ H' = _____

5. Reflect figure DFGH across the x-axis. Label vertices D", F", G", H" so that point D" is the reflection of D, F" is the reflection of F, G" is the reflection of G, and H" is the reflection of H.

D" = _____ G" = _____

F" = _____ H" = _____

6. Reflect figure DFGH across line s. Label vertices D'''F'''G'''H''' so that point D''' is the reflection of D, F''' corresponds to F, G''' to G, and H''' to H.

D''' = _____ G''' = _____

F''' = _____ H''' = _____

---

7. Reflect quadrilateral MNOP across the y-axis. Label vertices M'N'O'P' so that point M' is the reflection of point M, N' is the reflection of N, O' is the reflection of O, and P' is the reflection of P.

M' = _____ O' = _____

N' = _____ P' = _____

8. Reflect figure MNOP across the x-axis. Label vertices M", N", O", P" so that point M" is the reflection of M, N" is the reflection of N, O" is the reflection of O, and P" is the reflection of P.

M" = _____ O" = _____

N" = _____ P" = _____

9. Reflect figure MNOP across line w. Label vertices M'''N'''O'''P''' so that point M''' is the reflection of M, N''' corresponds to N, O''' to O, and P''' to P.

M''' = _____ O''' = _____

N''' = _____ P''' = _____

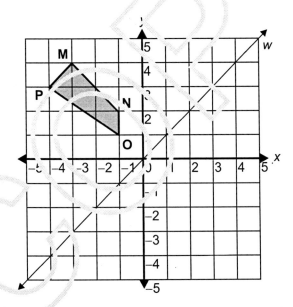

ROTATIONS

A **rotation** of a geometric figure shows motion around a point.

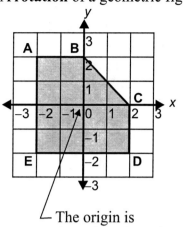

The origin is the point of rotation.

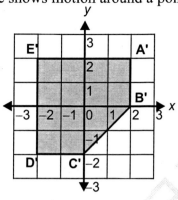

Figure ABCDE has been rotated $\frac{1}{4}$ of a turn clockwise around the origin to form A'B'C'D'E'.

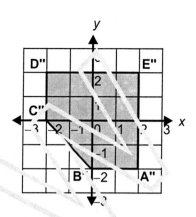

Figure ABCDE has been rotated $\frac{1}{2}$ of a turn around the origin to form A"B"C"D"E".

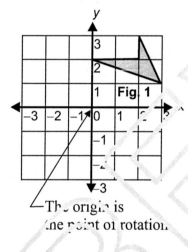

The origin is the point of rotation.

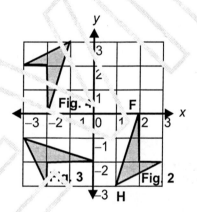

Figure 1 is rotated in $\frac{1}{4}$ turns around the origin. Figure 2 is a $\frac{1}{4}$ clockwise rotation of Figure 1. Figure 3 is a $\frac{1}{2}$ rotation of Figure 1. Figure 4 is a $\frac{3}{4}$ clockwise rotation of Figure 1.

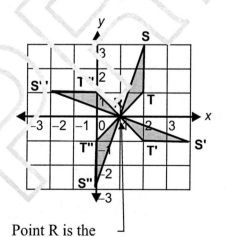

Triangle RST is rotated around point R. Triangle RS'T' is a $\frac{1}{4}$ clockwise rotation of triangle RST. Triangle RS"T" is a $\frac{1}{2}$ rotation of triangle RST. Triangle RS'''T''' is a $\frac{3}{4}$ clockwise rotation of triangle RST.

Point R is the point of rotation

Draw the following rotations, and record the new coordinates of the rotation. The figure for the first problem is drawn for you.

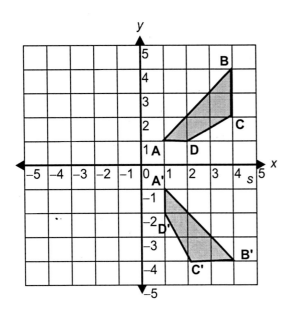

1. Rotate figure ABCD around the origin clockwise $\frac{1}{4}$ turn. Label the vertices A', B', C', and D' so that point A' corresponds to the rotation of point A, B' corresponds to B, C' to C, and D' to D.

   A' = _____   C' = _____

   B' = _____   D' = _____

2. Rotate figure ABCD around the origin clockwise $\frac{1}{2}$ turn. Label the vertices A", B", C", and D" so that point A" corresponds to the rotation of point A, B" corresponds to B, C" to C, and D" to D.

   A" = _____   C" = _____

   B" = _____   D" = _____

3. Rotate figure ABCD around the origin clockwise $\frac{3}{4}$ turn. Label the vertices A''', B''', C''', and D''' so that point A''' corresponds to the rotation of point A, B''' corresponds to B, C''' to C, and D''' to D

   A''' = _____   C''' = _____

   B''' = _____   D''' = _____

---

4. Rotate figure MNO around point O clockwise $\frac{1}{4}$ turn. Label the vertices M', N', and O so that point M' corresponds to the rotation of point M and N' corresponds to N.

   M' = _____   N' = _____

5. Rotate figure MNO around point O clockwise $\frac{1}{2}$ turn. Label the vertices M", N", and O so that point M" corresponds to the rotation of point M, and N" corresponds to N.

   M" = _____   N" = _____

6. Rotate figure MNO around point O clockwise $\frac{3}{4}$ turn. Label the vertices M''', N''', and O so that point M''' corresponds to the rotation of point M, and N''' corresponds to N.

   M''' = _____   N''' = _____

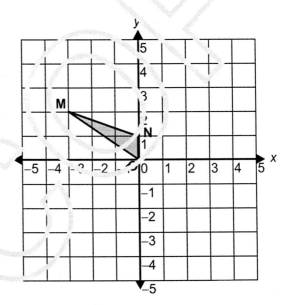

TRANSLATIONS

A **translation** of a geometric figure is a duplicate of the figure slid along a path.

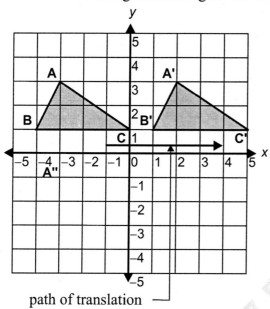

path of translation

Triangle A'B'C' is a translation of triangle ABC. Each point is translated 5 spaces to the right. In other words, the triangle slid 5 spaces to the right. Look at the path of translation. It gives the same information as above. Count the number of spaces across given by the path of translation, and you will see it represents a move 5 spaces to the right. Each new point is found at $(x + 5, y)$.

Point A is at $(-3, 3)$. Therefore, A' is found at $(-3 + 5, 3)$ or $(2, 3)$.

B is at $(-4, 1)$ so B' is at $(-4 + 5, 1)$ or $(1, 1)$.

C is at $(0, 1)$ so C' is at $(0 + 5, 1)$ or $(5, 1)$.

Quadrilateral FGHI is translated 5 spaces to the right and 3 spaces down. The path of translation shows the same information. It points right 5 spaces and down 3 spaces. Each new point is found at $(x + 5, y - 3)$.

Point F is located at $(-4, 3)$. Point F' is located at $(-4 + 5, 3 - 3)$ or $(1, 0)$.

Point G is at $(-2, 5)$. Point G' is at $(-2 + 5, 5 - 3)$ or $(3, 2)$.

Point H is at $(-1, 4)$. Point H' is at $(-1 + 5, 4 - 3)$ or $(4, 1)$.

Point I is at $(-1, 2)$. Point I' is at $(-1 + 5, 2 - 3)$ or $(4, -1)$.

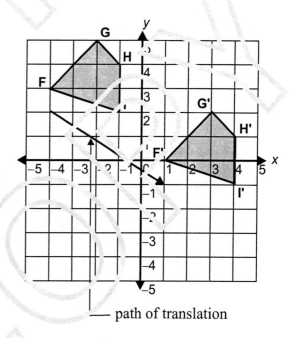

path of translation

Draw the following translations and record the new coordinates of the translation. The figure for the first problem is drawn for you.

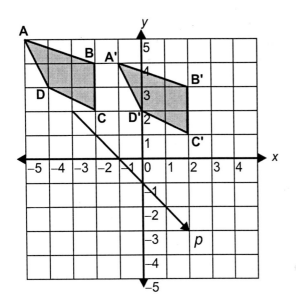

1. Translate figure ABCD 4 spaces to the right and 1 space down. Label the vertices of the translated figure A', B', C', and D' so that point A' corresponds to the translation of point A, B' corresponds to B, C' to C, and D' to D.

   A' = _____   C' = _____

   B' = _____   D' = _____

2. Translate figure ABCD 5 spaces down. Label the vertices of the translated figure A", B", C", and D" so that point A" corresponds to the translation of point A, B" corresponds to B, C" to C, and D" to D.

   A" = _____   C" = _____

   B" = _____   D" = _____

3. Translate figure ABCD along the path of translation, p. Label the vertices of the translated figure A''', B''', C''', and D''' so that point A''' corresponds to the translation of point A, B''' corresponds to B, C''' to C, and D''' to D.

   A''' = _____   C''' = _____

   B''' = _____   D''' = _____

---

4. Translate triangle FGH 6 spaces to the left and 3 spaces up. Label the vertices of the translated figure F', G', and H' so that point F' corresponds to the translation of point F, G' corresponds to G, and H' to H.

   F' = _____   G' = _____   H' = _____

5. Translate triangle FGH 4 spaces up and 1 space to the left. Label the vertices of the translated triangle F"G"H" so that point F" corresponds to the translation of point F, G" corresponds to G, and H" to H.

   F" = _____   G" = _____   H" = _____

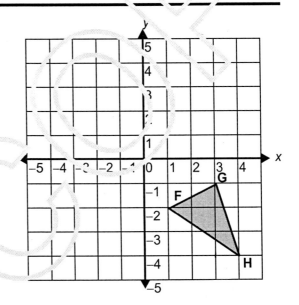

296 Copyright © American Book Company

TRANSFORMATION PRACTICE

Answer the following questions regarding transformations.

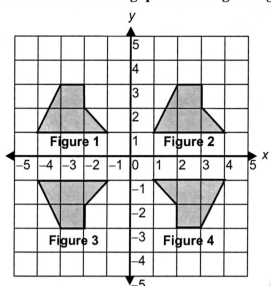

1. Which figure is a rotation of Figure 1? _____

   How far is it rotated? _____

2. Which figure is a translation of Figure 1? _____

   How far and in which direction(s) is it translated?
   _____

3. Which figure is a reflection of Figure 1? _____

4. Translate quadrilateral ABCD so that point A' which corresponds to point A is located at coordinates (−4, 3). Label the other vertices B' to correspond to B, C' to C, and D' to D. What are the coordinates of B', C', and D'?

   A' = _____ C' = _____

   B' = _____ D' = _____

5. Reflect quadrilateral ABCD across line m. Label the coordinates A", B", C", D" so that point A" corresponds to the reflection of point A, B" corresponds to the reflection of B, and C" corresponds to the reflection of C. What are the coordinates of A", B", C", and D"?

   A" = _____ C" = _____

   B" = _____ D" = _____

6. Rotate quadrilateral ABCD ¼ turn counterclockwise around point D. Label the points A'''B'''C'''D''' so that A''' corresponds to the rotation of point A, B''' corresponds to B, C''' to C, and D''' to D. What are the coordinates of A''', B''', C''' and D'''?

   A''' = _____ C''' = _____

   B''' = _____ D''' = _____

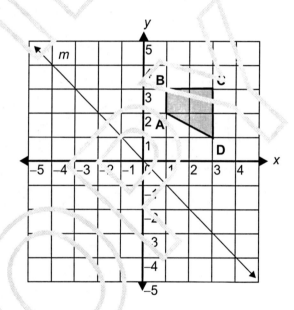

Copyright © American Book Company

SYMMETRY

Many geometric figures are symmetrical or have **symmetry**. Geometric figures can have three types of symmetry: **reflectional**, **rotational**, and **translational**.

REFLECTIONAL SYMMETRY

A figure has **reflectional symmetry** if you can draw a line through the figure that divides it into two mirror images. The mirror image line is called the **line of symmetry**. Look at the figures below.

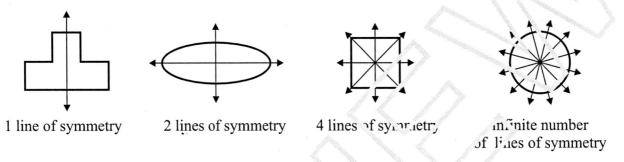

1 line of symmetry   2 lines of symmetry   4 lines of symmetry   infinite number of lines of symmetry

ROTATION SYMMETRY

A figure has **rotational symmetry** if the image will lie on top of itself when rotated through some angle other than $360°$. Look at the figures below.

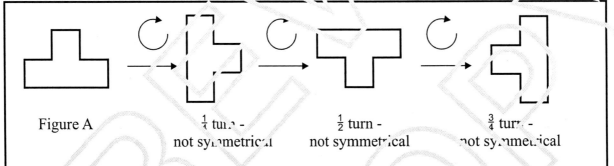

Figure A $\frac{1}{4}$ turn - not symmetrical $\frac{1}{2}$ turn - not symmetrical $\frac{3}{4}$ turn - not symmetrical

Figure A does not have rotational symmetry. It cannot be rotated so that the rotated image falls exactly on top of the original image.

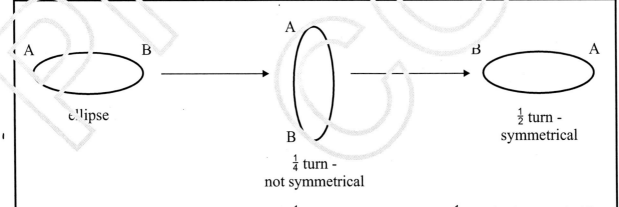

ellipse $\frac{1}{4}$ turn - not symmetrical $\frac{1}{2}$ turn - symmetrical

An **ellipse** has rotational symmetry at each $\frac{1}{2}$ turn. An ellipse rotated $\frac{1}{2}$ turn looks exactly like the original ellipse. All the points would lie exactly on top of each other.

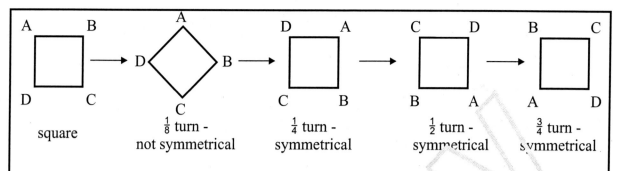

A **square** has rotational symmetry at each $\frac{1}{4}$ turn. A square rotated $\frac{1}{4}$, $\frac{1}{2}$, or $\frac{3}{4}$ looks identical to the original image.

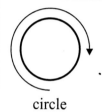

A **circle** has complete rotational symmetry. No matter how much you rotate a circle, the rotated image will always look identical to the original circle.

TRANSLATIONAL SYMMETRY

A geometric pattern has **translational symmetry** if an image can be slid a fixed distance in opposite directions to obtain the same pattern.

This pattern has translational symmetry because when you slide it **horizontally**, it matches the same pattern.

This pattern has translational symmetry because when you slide it **vertically**, it matches the same pattern.

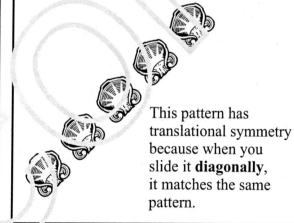

This pattern has translational symmetry because when you slide it **diagonally**, it matches the same pattern.

This pattern **does not** have translational symmetry. The image does not match the same pattern as you slide it in any direction.

SYMMETRY PRACTICE

Match each figure below to the letter that describes its symmetry. Some have more than one answer. Choose all the letters that apply.

1. _____

2. _____

3. _____

4. _____

5. _____

6. _____

7. _____

8. _____

9. _____

10. _____

A. $\frac{1}{4}$ turn rotational symmetry

B. $\frac{1}{2}$ turn rotational symmetry

C. complete rotational symmetry

D. reflectional symmetry

E. translational symmetry

F. not symmetrical

11. How many lines of symmetry can be drawn though a regular pentagon?

    _____

12. How many lines of symmetry can be drawn through the following parallelogram?

    _____

CHAPTER REVIEW

1.

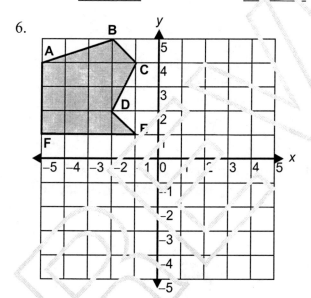

Draw the reflection of image ABCD over the y-axis. Label the points A', B', C', and D'. List the coordinates of these points below.

2. A' _____   4. C' _____

3. B' _____   5. D' _____

6.

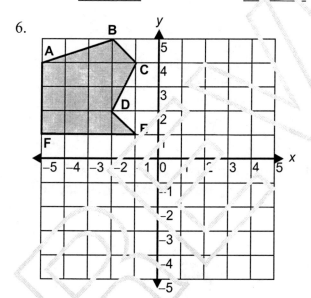

Rotate the figure above a $\frac{1}{2}$ turn about the origin, O. Label the points A', B', C', D', E', and F'. List the coordinates of these points below.

7. A' _____   10. D' _____

8. B' _____   11. E' _____

9. C' _____   12. F' _____

13.

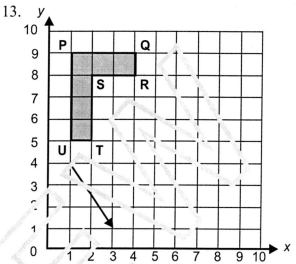

Use the translation described by the arrow to translate the polygon above. Label the points P', Q', R', S', T', and U'. List the coordinates of each.

14. P' _____   17. S' _____

15. Q' _____   18. T' _____

16. R' _____   19. U' _____

Record the kind(s) of symmetry, if any, shown in each example below.

20.    _____

21.    _____

22.    _____

23.    _____

CHAPTER TEST

1.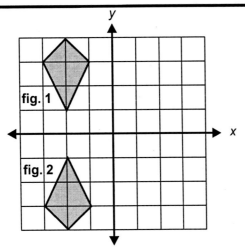

   Figure 1 goes through a transformation to form figure 2. Which of the following descriptions fits the transformation shown?

   A. reflection across the x-axis
   B. reflection across the y-axis
   C. $\frac{3}{4}$ clockwise rotation around the origin
   D. translation down 2 units

2.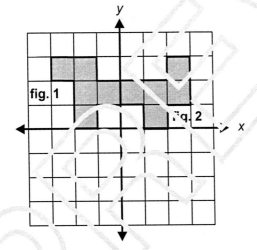

   Figure 1 goes through a transformation to form figure 2. Which of the following descriptions fits the transformation shown?

   A. reflection across the x-axis
   B. reflection across the y-axis
   C. $\frac{1}{4}$ clockwise rotation around the origin
   D. translation right 3 units

3.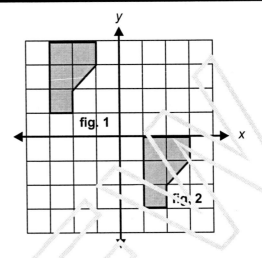

   Figure 1 goes through a transformation to form figure 2. Which of the following descriptions fits the transformation shown?

   A. reflection across the x-axis
   B. reflection across the origin
   C. $\frac{1}{2}$ clockwise rotation around the origin
   D. translation right 4 units and down 4 units

4. What kind(s) of symmetry does the following figure have? Choose the best answer.

   I. reflectional symmetry
   II. $\frac{1}{4}$ rotational symmetry
   III. translational symmetry

   A. I
   B. II
   C. I and II
   D. III

5. What kind(s) of symmetry does the following figure have? Choose the best answer.

A. only reflectional symmetry
B. only translational symmetry
C. reflectional and translational symmetry
D. no symmetry

6. What kind(s) of symmetry does the following figure have? Choose the best answer.

A. reflectional symmetry
B. rotational symmetry
C. reflectional and rotational symmetry
D. no symmetry

7. What kind(s) of symmetry does the following figure have? Choose the best answer.

A. $\frac{1}{4}$ rotational symmetry
B. $\frac{1}{2}$ rotational symmetry
C. complete rotational symmetry
D. no symmetry

8. What kind(s) of symmetry does the following figure have? Choose the best answer.

A. $\frac{1}{2}$ rotational symmetry
B. reflectional symmetry
C. translational symmetry
D. no symmetry

9.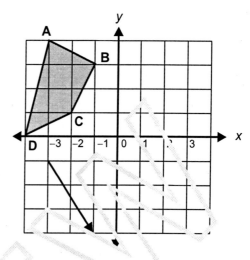

If the figure above is translated in the direction described by the arrow, what will be the new coordinates of point B after the transformation?

A. (2, 1)
B. (0, 1)
C. (1, 0)
D. (1, 1)

10. How many lines of symmetry can be drawn through the following figure? Choose the best answer.

A. 2
B. 4
C. infinite
D. none

11. How many lines of symmetry can be drawn through the following figure? Choose the best answer.

A. 2
B. 4
C. infinite
D. none

FORMULA SHEET

Formulas that you may need to answer questions on this exam are found below.

Area of a square = s^2

Area of a rectangle = lw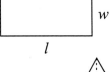

Area of a triangle = $\frac{1}{2} bh$

Area of a circle = πr^2

Circumference = πd or $2\pi r$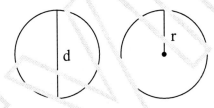

π = Pi = 3.14 or $\frac{22}{7}$

Area of a trapezoid = $\dfrac{(b_1 + b_2) \times h}{2}$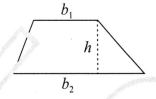

Volume of a rectangular prism = $l \times w \times h$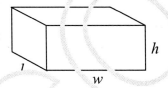

Volume of a cylinder = $\pi r^2 h$

Pythagorean Theorem: $a^2 + b^2 = c^2$

Post Test For Basics Made Easy Mathematics

1. In the Venn diagram below, how many members are in the following set?

    {brother} ∩ {sister}

    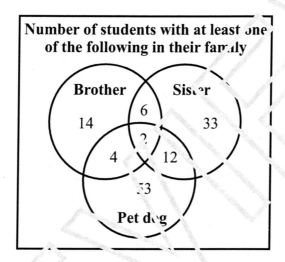

    A. 6
    B. 8
    C. 53
    D. 55

2. Which of the following statements is true?

    A. {1, 2, 4, 5} ∪ ∅ = ∅
    B. {c} ∉ {c, a, t}
    C. {1, g} ⊂ {f, g, h, i}
    D. {k, l, m} ∪ ∅ = ∅

3. Which of the following numbers is prime?

    A. 9
    B. 11
    C. 12
    D. 15

4. Which is a reasonable estimate of the answer?
$$16,729 \div 100$$

   A. 16,700
   B. 1,670
   C. 167
   D. 17

5. Jed has 155 head of cattle. Each eats 31 pounds of silage every day. How much silage does Jed feed his cattle every day?

   A. 5 lb
   B. 3705 lb
   C. 4705 lb
   D. 4805 lb

6. Which fraction is greater than $\frac{3}{5}$?

   A. $\frac{7}{8}$
   B. $\frac{1}{3}$
   C. $\frac{4}{15}$
   D. $\frac{3}{8}$

7. Jerome bought $1\frac{1}{2}$ pounds of jelly beans, $\frac{2}{8}$ pound of gum balls, and $2\frac{3}{4}$ pounds of chocolate. How many total pounds of candy did he buy?

   A. $3\frac{7}{8}$ pounds
   B. $4\frac{3}{4}$ pounds
   C. $4\frac{7}{8}$ pounds
   D. 5 pounds

8. Which set of decimals is in order from LEAST to GREATEST?

   A. .64, .604, .064, .06
   B. .603, .75, .08, .098
   C. .5, .064, .701, .901
   D. .007, .069, .69, .7

9. Which is the best buy?

    A. 3 candy bars for $0.96
    B. 4 candy bars for $1.00
    C. 5 candy bars for $1.20
    D. 6 candy bars for $2.10

10. Monica's monthly salary is $2,100. Her deductions are $157.50 for FICA, $302 for federal income tax, and $57.80 for state income tax. What is her take-home pay?

    A. $1,682.70
    B. $1,617.30
    C. $1,582.69
    D. $1,582.70

11. On a blueprint of a house, the scale is 0.25 inches equals 2 feet. How wide is the kitchen if it measures 1.5 inches on the blueprint?

    A. $0.\overline{33}$ feet
    B. 3 feet
    C. 12 feet
    D. 15 feet

12. A pine tree casts a shadow 9 feet long. At the same time, a rod measuring 4 feet casts a shadow 1.5 feet long. How tall is the pine tree?

    A. 3.375 feet
    B. 13.5 feet
    C. 24 feet
    D. 54 feet

13. 122% written as a fraction is

    A. $1\frac{11}{50}$
    B. $2\frac{3}{25}$
    C. $12\frac{1}{5}$
    D. 122

14. Tate is saving his money to buy a go-cart. The go-cart he wants costs $240. His mother agreed to chip in $60. What percent of the total cost of the go-cart is his mother contributing?

    A. 20%
    B. 25%
    C. 40%
    D. 75%

15. Mark invested $1,340 in a simple interest account at 3.5% interest. How much interest did he earn after 4 years?

    A. $ 46.90
    B. $ 187.60
    C. $ 469.00
    D. $1,876.00

16. Emily had $100.00 to spend. She spent $37.85 on clothes. How much change should she receive from $40.00?

    A. $ 2.15
    B. $ 3.15
    C. $60.00
    D. $62.15

17. Lee works part time at a fast-food restaurant and makes $45 per week. He saves 75% towards college. How much will he have saved after 32 weeks?

    A. $1080.00
    B. $ 33.75
    C. $ 53.33
    D. Not enough information is given.

18. $2354^0 =$

    A. 0
    B. 1
    C. 2354
    D. 23540

19. Which unit of measure would be most appropriate to measure the dimensions of a classroom?

    A. millimeters
    B. centimeters
    C. meters
    D. kilometers

20. Which of the following is equal to 5 L?

    A. 0.005 mL
    B. 0.5 kL
    C. 5,000 mL
    D. 5,000 kL

21.
    1 gal 1 pt
    ×   5

    A. 5 gal 1 qt 1 pt
    B. 5 gal 2 qt 1 pt
    C. 6 gal 1 qt 1 pt
    D. 6 gal   1 pt

22.
    5 yd 1 ft 3 in
    − 2 yd 2 ft 6 in

    A. 2 yd 1 ft 9 in
    B. 2 yd 1 ft 6 in
    C. 3 yd 2 ft 3 in
    D. 8 yd 9 in

23. The Thompson family wants to make a graph to show the percent of money spent on vacation in the following categories:

    entertainment   5%
    hotel   55%
    souvenirs   8%
    food   20%
    gasoline   12%

    What is the best kind of graph to use to display their results?

    A. bar graph
    B. single line graph
    C. circle graph
    D. multiple line graph

24. What would be the total cost of 18 pairs of Sports Socks and 12 pairs of Anklets?

SOCKS-A-PLENTY WAREHOUSE		
Item	3 pair price	6 pair price
Anklets	$4.50	$ 8.50
Sport Socks	$5.50	$11.50
Support Socks	$9.00	$15.75
Knee Hi	$6.00	$11.25
Please add 10% for shipping and handling		

    A. $56.65
    B. $51.50
    C. $57.00
    D. $339.90

25. The Kreise family drove from Miami to West Palm Beach, and then they drove from West Palm Beach to Orlando. How many miles did they travel total?

	Jacksonville	Miami	Orlando	Tallahassee	Tampa	West Palm Beach
Jacksonville	0	341	142	163	202	277
Miami	341	0	229	478	273	66
Orlando	142	229	0	257	84	165
Tallahassee	163	478	257	0	275	414
Tampa	202	273	84	275	0	226
West Palm Beach	277	66	165	414	226	0

A. 66
B. 165
C. 202
D. 231

26. The student council surveyed the student body on favorite lunch items. The frequency chart below shows the results of the survey.

Favorite Lunch Item	Frequency
corndog	140
hamburger	245
hotdog	210
pizza	235
spaghetti	90
other	65

Which lunch item indicates the mode of the data?

A. other
B. hotdog
C. hamburger
D. corndog

27. Which of the following sets of numbers has a range of 73?

A. { 29, 19, 72, 68, 39 }
B. { 81, 85, 37, 41, 60 }
C. { 17, 12, 9, 47, 82 }
D. { 62, 86, 44, 78, 95 }

28. $6 + 18 \times 4^2 - 4 =$

   A. 146
   B. 188
   C. 290
   D. 380

29. Which of the following shows the Identity Property of Addition?

   A. $a + 0 = a$
   B. $a + b = b + a$
   C. $a + (-a) = 0$
   D. $a + (b + c) = (a + b) + c$

30. $-|-8| - |-2| =$

   A. 10
   B. −10
   C. 6
   D. −6

31. Which of the following algebraic expressions corresponds to:

   "the product of 5 and x divided by 3 fewer than y"

   A. $\dfrac{5x}{y-3}$
   B. $\dfrac{5}{x(y-3)}$
   C. $\dfrac{5x}{3-y}$
   D. $\dfrac{5x}{y+3}$

32. In the following problem, $a = -2$, $b = 4$ and $c = 3$. Solve the problem.

   $5 + 3a - 4c =$

   A. 14
   B. −17
   C. −49
   D. −180

33. If $\frac{n}{6} = -8$, what is *n*?

    A. −48
    B. 14
    C. 8
    D. −2

34. Find *v*: $v + 3 \geq 24$

    A. $v \leq 21$
    B. $v \leq 27$
    C. $v \geq 21$
    D. $v \geq 27$

35. Which number line below represents the following inequality?

    $-2 < x \leq 3$

    A. number line with open circle at −2 and closed circle at 3, shaded between
    B. number line with closed circle at −2 and open circle at 3, shaded outside
    C. number line with open circle at −2 and closed circle at 3, shaded to the right
    D. number line with closed circle at −2 and open circle at 3, shaded to the right

36. Andrea has 7 teddy bears in a row on a shelf in her room. How many ways can she arrange the bears in a row on her shelf?

    A. 7
    B. 49
    C. 343
    D. 5,040

37. The buffet line offers 5 kinds of meat, 3 different salads, a choice of 4 desserts, and 5 different drinks. If you chose one food from each category, from how many combinations would you have to choose?

    A. 17
    B. 20
    C. 60
    D. 300

38. What is the measure of ∠ BRF?

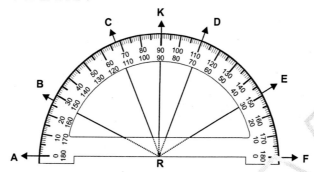

   A. 27°
   B. 33°
   C. 153°
   D. 167°

39. Name the angle that is vertical to ∠ EAF

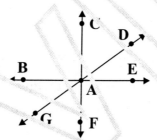

   A. ∠ EAD
   B. ∠ CAB
   C. ∠ BAG
   D. ∠ GAF

40. Which two angles are supplementary?

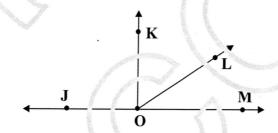

   A. ∠ JOL and ∠ KOM
   B. ∠ LOM and ∠ KOL
   C. ∠ JOK and ∠ KOL
   D. ∠ KOM and ∠ JOK

41. The following triangles are similar.

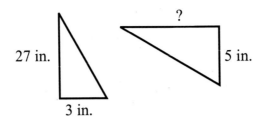

What is the length of the missing side?

A. 45 in.
B. 18 in.
C. 15 in.
D. 10 in.

42.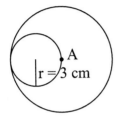

Point A is in the center of the larger circle. The radius of the smaller circle is 3 cm. What is the area of the larger circle? Use 3.14 for π. Area = πr^2.

A. 9.42
B. 18.84
C. 28.26
D. 113.04

43. Find the volume of the following pyramid.

Use the formula $v = \frac{lwh}{3}$

l = length
w = width
h = height

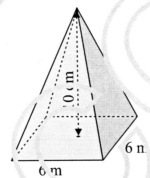

A. 22 cm³
B. 120 cm³
C. 360 cm³
D. 1,080 cm³

44. What is the surface area of a cylinder with an 8 inch diameter and a height of 3 inches?

    Use the formula $SA = 2\pi r^2 + 2\pi rh$
    Use $\pi = 3.14$

    A. 100.48 in^2
    B. 50.24 in^2
    C. 75.36 in^2
    D. 175.84 in^2

45. Jack is going to paint the ceiling and four walls of a room that is 10 feet wide, 12 feet long, and 10 feet from floor to ceiling. How many square feet will he paint?

    A. 120 square feet
    B. 560 square feet
    C. 680 square feet
    D. 1,200 square feet

46.
    ```
    ←—•————+—•————+—•————+—•—→
      7 P   8 Q   9 R   10 S
    ```

    Where is $\sqrt{83}$ located on the number line above?

    A. Point P
    B. Point Q
    C. Point R
    D. Point S

47.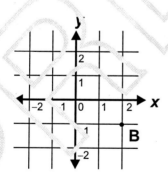

    What are the coordinates of point B above?

    A. (2, −1)
    B. (−1, 2)
    C. (−2, 1)
    D. (1, −2)

48.

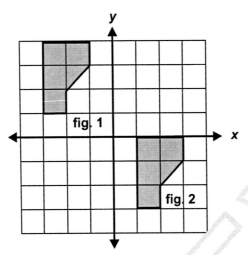

Figure 1 goes through a transformation to form figure 2. Which of the following descriptions fits the transformation shown?

A. reflection across the x-axis
B. reflection across the origin
C. $\frac{1}{2}$ clockwise rotation around the origin
D. translation right 4 units and down 4 units

49.

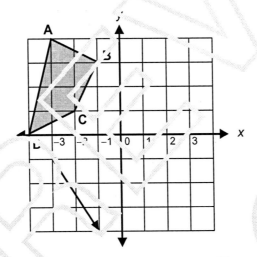

If the figure above is translated in the direction described by the arrow, what will be the new coordinates of point B after the transformation?

A. (2, 1)
B. (0, 1)
C. (1, 0)
D. (1, 1)

50. How many lines of symmetry can be drawn through the following figure? Choose the best answer.

    A. 2
    B. 4
    C. infinite
    D. none